The Isle of Wight

LANDSCAPE AND GEOLOGY

John Downes

The Isle of Wight

LANDSCAPE AND GEOLOGY

John Downes

THE CROWOOD PRESS

First published in 2021 by
The Crowood Press Ltd
Ramsbury, Marlborough
Wiltshire SN8 2HR

enquiries@crowood.com

www.crowood.com

British Library Cataloguing-in-Publication Data
A catalogue record for this book is available from the British Library.

ISBN 978 1 78500 892 4

Acknowledgements
I wish to acknowledge the invaluable contribution of photographic images by Antoinette Pearson of the Open University (pp. 9, 13, 15, 16T,B, 17B, 21, 24, 25all, 26all, 27T, 28all, 30, 35T,B, 36, 37B, 38-39B, 41R, 47T, 52T,B, 53T,B, 55, 77, 80B, 103T, 105T), Stephen Hallett of Cranfield University (pp. 34, 38T, 39T, 45T, 46T,M, 49T, 61, 66, 73, 87, 90, 91T,B, 100, 104, 105B), Tony Cross of the Open University (pp. 37T, 45B, 47B, 82B, 83T,B, 102), Tony Waltham Geophotos (pp. 2, 31, 32T,B, 46B, 79T,B, 81, 96, 103B), Dinosaur Isle Museum (p.19T,B) and John Faulkner (p.94).

Drawings on pp. 40T, 40B and 41L are extracted with permission from *British Mesozoic Fossils* and *British Cenozoic Fossils*, British Museum (Natural History), 1975.

Images used under Creative Commons Licence are by Graham Horn (p.74), Peter Trimming (pp. 76, 99), Christine Matthews (p. 88), Ron Saunders (pp. 97T,B) and Mypix (pp. 67, 93, 98).

All other photographs and drawings are by the author.

Front cover: The chalk cliff at the western tip of the Needles headland.
Frontispiece: The outermost sea stacks of The Needles.

Typeset by Tony Waltham
Cover design by Blue Sunflower Creative
Printed and bound in India by Replika Press Pvt. Ltd.

Contents

Preface

The basic concept underpinning this book is the relationship between geology, the landscape and land use. It is written for the reader with an interest in seeing and understanding the geology and the physical and human geography of the Isle of Wight. Amateur geologists, students, hikers, holidaymakers and even the apocryphal 'man on the Clapham omnibus' should find this guide helpful in explaining the development of the island that the Romans called Vectis.

The wonderful coastline, which is some 110km long, exposes rocks that range in age from the Lower Cretaceous to Oligocene. The first six chapters of this book relate the geological story of how the rocks have been formed and what past environments they represent. Each chapter has relevant itineraries for those who want to try 'hands on' geology in the field, with essential information regarding directions, access and parking. In addition, there are sections that provide the reader with additional background material.

Landscape and scenery are covered in succeeding chapters, and this is followed by the story of human settlement from Neolithic times to the present day. Indeed, the impact of modern economic activities – dominated by the tourism – represents a considerable threat to the physical environment that we need to thoroughly understand in order to be able to protect and preserve the beautiful landscape of this island.

Most of the localities have been researched and recorded in academic publications (*see* Further Reading) but here I have attempted to bring geology alive to the enthusiastic amateur observer by concentrating on key features of the rocks and landscape. Indeed, being able to practise geology in the field, in an area such as the Isle of Wight, is one of the most rewarding and enjoyable experiences of life, providing a glimpse of the treasures of the Earth, often in the company of the 'salt of the earth' … your fellow explorers. Details of access to each locality are correct at the time of writing but you should be aware that ownership of land changes, and rights of way may be diverted or restricted from time to time, particularly where coastal erosion is undermining clifftop paths and collapsing the sides of chines. You may find the local bus services useful when planning walks. These are operated by Southern Vectis Omnibus Company, which runs a variety of tourist buses including the Island Coaster and the Needles Breezer during the summer season. Timetables and information on concessionary fares and holiday rover tickets can be found on the website www.islandbuses.info.

You should read the Geological Fieldwork Code, printed at the start of this book. It is designed to ensure both your safety and the preservation of the geological environment. Some localities are Regionally Important Geodiversity Sites (RIGS) and some are Sites of Special Scientific Interest (SSSI). There is also the Isle of Wight Local Geodiversity Action Plan (LGAP), the primary function of which is to create a strategy to promote the island through the conservation and sustainable development of its Earth heritage. Other administrative bodies, such as the National Trust, English Heritage, the Forestry Commission, the Wildlife Trust and the local authority, also play an important role in protecting and conserving sites of geological and biological interest. Furthermore, almost half of the Isle of Wight is designated as an

Area of Outstanding Natural beauty (AONB) in order to conserve and enhance its cultural heritage, wildlife and landscape.

For many years I have tutored courses for The Open University, where the contributions of Earth Science students from all walks of life have so often enriched arcane discussions concerning the origin of the rocks. It is in this spirit of enquiry that we need to approach geology in the field. Moreover, it is important to recognize that more often than not, there is no neat and tidy answer to the riddle of the rocks. Even the experts often disagree and rarely commit themselves fully one way or another; some preface their answers with expressions such as 'on the one hand there is the possibility that … while on the other it may be that…' In fact, it is often the case that more than one event in the past has produced what we see today.

Perhaps our best guide must still be the principle of uniformitarianism as proposed by the Scottish philosopher James Hutton in the eighteenth century. His thesis was that the present is the key to the past; in other words, such contemporary activity as erosion by rivers or glaciers, the deposition of sands on a beach or the eruption of volcanoes involves processes that have operated for millions of years in the making of our Earth. Hutton succinctly expressed his thoughts on the origins of the rocks when he declared, 'There is no vestige of a beginning … no prospect of an end'.

As to the subject of this book; we can do no better than read the words of the Reverend J. Cecil Hughes, who spent many years on the island studying its rocks. In 1922 he wrote in the preface to his book *The Geological Story of the Isle of Wight* (reprinted in 2011):

No better district could be chosen to begin the study of geology than the Isle of Wight. The splendid coastal sections all round its shores, the variety of strata within so small an area, the great interest of those strata, the white chalk cliffs and coloured sands, the abundant and interesting fossils to be found in the rocks, awaken … a desire to know something of the story written in the rocks.

Geological Fieldwork Code

1. Follow the Country Code and observe local bye-laws. Remember to shut gates and to leave no litter.
2. Avoid undue disturbance to farm animals, wildlife or natural vegetation. Keep clear of nesting birds, particularly on coastal cliffs.
3. Always seek permission before entering private land.
4. Observe and record; make field sketches and photograph geological sections, but do not damage the outcrop by hammering indiscriminately.
5. Restrict the collecting of specimens to a minimum and do not remove *in situ* specimens.
6. Leave the site in as good a condition as possible for those who come after you.
7. Beware of dangerous cliffs and rock faces, taking care not to dislodge loose rocks.
8. In coastal localities, make sure that you consult tide timetables and warning notices. It is always best to work on a falling tide.
9. When visiting isolated places, make sure that someone knows where you are going and your estimated time of return. It is best to go with a companion rather than alone.
10. Always wear suitable footwear and outdoor clothing. Wear a safety helmet where advisable. Carry a mobile phone, but remember that it may fail to get a signal below cliffs or in remote country.

A more complete code is published by The Geologists' Association.

Safety Notice

It is the responsibility of the user of this field guide to take all necessary precautions, particularly when on exposed cliff sections where falling rocks and adverse tidal conditions can create dangerous situations. Check tide timetables before beginning a coastal excursion to avoid being trapped by the incoming tide (see *Time and Tide on the Foreshore* in Chapter 8).

CHAPTER 1

Introduction

The Isle of Wight has been described as the 'diamond in Britain's geological crown', and it does indeed provide a wonderful example of how geology influences the development of the landscape and the imprint of human activity on the land. Moreover, the island has been studied and written about in detail for several centuries. The scientific philosopher Robert Hooke (1635–1703) was born in Freshwater on the island and was a leading scientist of his day, using the newly invented reflecting microscope to study fossils, plant and insects, which he used as illustrations in his book *Micrographia*.

By Victorian times, scientific interest in the island had increased, as fossil bones and other evidence of past life emerged. Gideon Mantell (1790–1852) wrote several papers on the dinosaur remains found in the Wealden Beds of Compton Bay. Charles Lyell, the famous Scottish geologist, made use of his work on the island's strata in writing his influential *Principles of Geology* (1830), which Darwin took with him on his scientific expedition in the *Beagle*. Well-known literary figures including Tennyson, Keats and Walter Scott were also attracted to the picturesque beauty of the Isle of Wight, and when Queen Victoria established her summer palace at Osborne in 1851, the island's popularity was assured. Iconic views of the chalk cliffs of the Needles, the coloured cliffs of Alum Bay and the holiday beaches of Shanklin, so beloved of the railway artists of the 1930s, helped fuel the tourist industry that today provides the lifeblood of the island's economy.

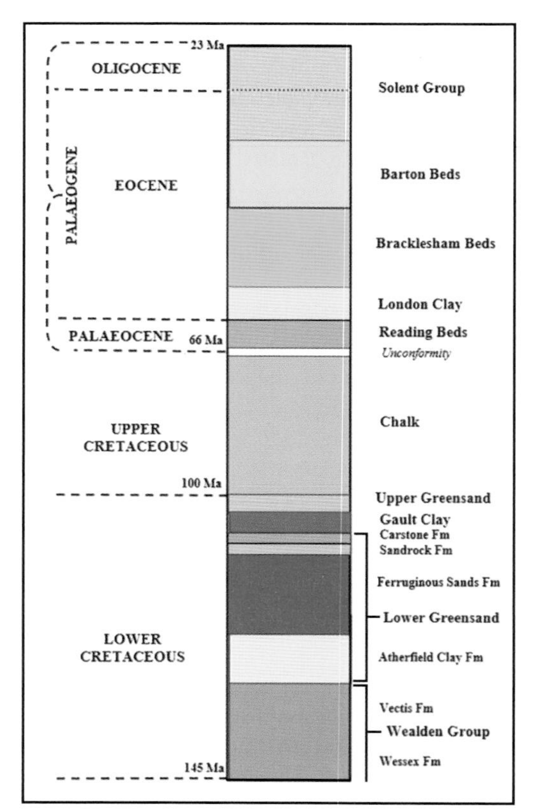

Time chart showing the succession of bedrocks exposed at the surface across the Isle of Wight.

A Geological Timescale

To explain the geology of the Isle of Wight, we need to consider how it fits into the framework of geological time. The oldest rocks exposed at the surface are of Lower Cretaceous age (140–100 million years, or Ma); they include the Wealden Beds at the base, followed by the Lower Greensand, Gault Clay and Upper Greensand. Compared to the Cambrian rocks of Wales that are over 500 million years old, the strata on the Isle of Wight are relatively young! The chalk is of Upper Cretaceous age (100–66 Ma) at the top of which is a break in the sequence.

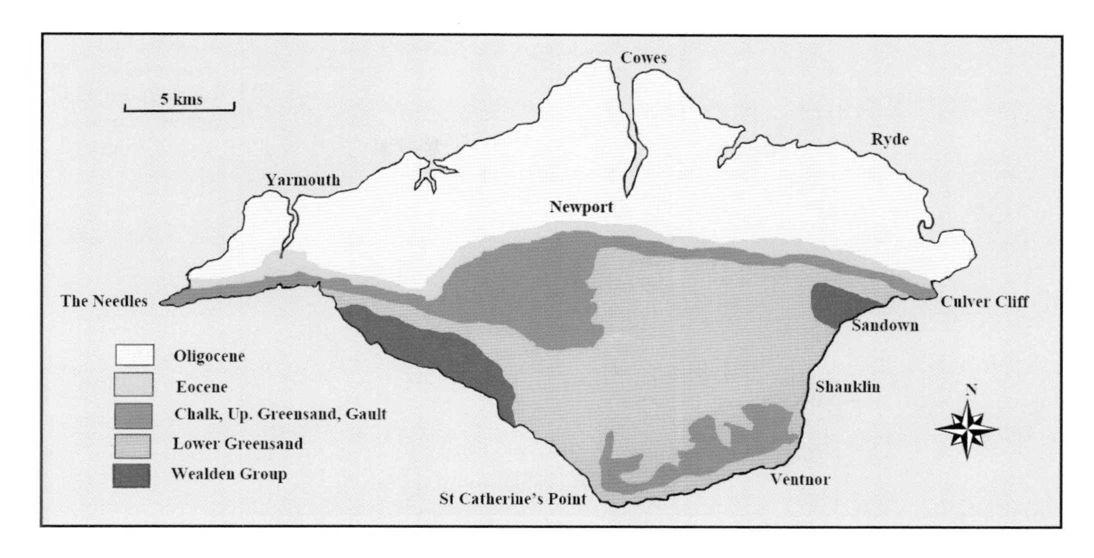

Simplified geological map of the Isle of Wight.

This is because at the end of the Cretaceous the whole area was uplifted and eroded before the succeeding Paleogene rocks were deposited. We refer to this hiatus in the succession as an unconformity, and in some areas the time interval may be measured in tens of millions of years. The youngest 'solid' rocks on the island belong to the Oligocene period (34–23 Ma); after this there is no material evidence on the ground except for scattered superficial deposits such as plateau gravels, solifluction deposits, river alluvium and clay-with-flints covering the chalk downlands. However, although there are no rocks of Miocene age (23–5 Ma), this was a time of great upheaval in southern England and the Isle of Wight when the Alpine earth movements uplifted, folded and faulted the pre-existing rocks.

The most significant Alpine structure is the east–west trending monocline forming the chalk ridge extending from the Needles in the west to Culver Cliff in the east. This fold has a steep northerly dip and an almost horizontal southern flank as a result of pressure from the south. As you will see from the structural map of the Isle of Wight, the fold axis is in two sections: from the Needles, the western section runs ESE towards Shorwell, and the eastern section extends

The Ferruginous Sands at Red Cliff looking towards the white chalk exposed along the cliffs of Culver Down.

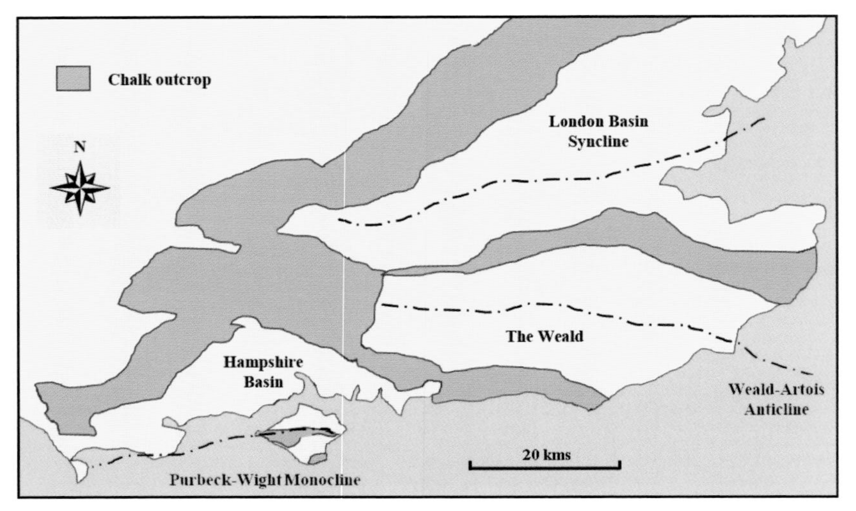

Miocene fold structures in southern England produced by the Alpine orogeny.

roughly west to east from Newport to Culver Cliff. On the east side of Freshwater Bay, you can see not only monoclinal folding in the chalk but also numerous shear planes cutting across the bedding. These are caused by the intense compressional forces applied to the rocks during folding. The axial trend of the monocline can be traced westwards across to the Isle of Purbeck in Dorset. The fold was continuous until breached by the sea when the Isle of Wight was separated from the mainland by the Flandrian rise in sea level, some 12,000 years ago.

There are several other important folds on the island. The Brighstone anticline runs WNW–ESE through the Wealden Beds and Lower Greensand around Brighstone Bay but much of its southwestern flank has now been eroded by the sea. On the opposite side of the island, the axis of the Sandown anticline trends NW–SE and exposes Wealden Beds in its core. North of the chalk ridge, the Bembridge syncline extends from Foreland Point ESE–WNW across the island. The Bembridge Limestone forms the flanks of the syncline while the younger Hamstead Beds occupy the central axial zone. This is an asymmetrical fold since the beds on the southern side are much steeper than those to the north. In the northwest of the island, the Bouldnor syncline and the complementary Porchfield anticline and Thorness syncline extend across the Solent to the mainland.

Landscape and Geological Structure

The topography of the Isle of Wight is closely related to the underlying geological outcrops. It is the east–west chalk ridge that forms the long axis of the lozenge-shaped island, which it divides into two areas of very different character. The central chalk outcrop in both its eastern and western extremities is relatively

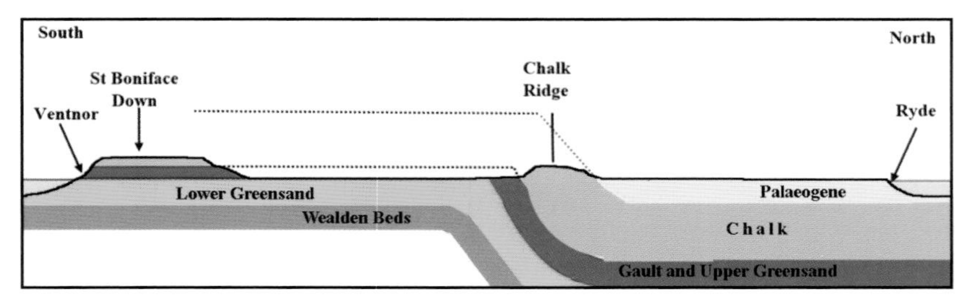

Geological section across the Isle of Wight monocline.

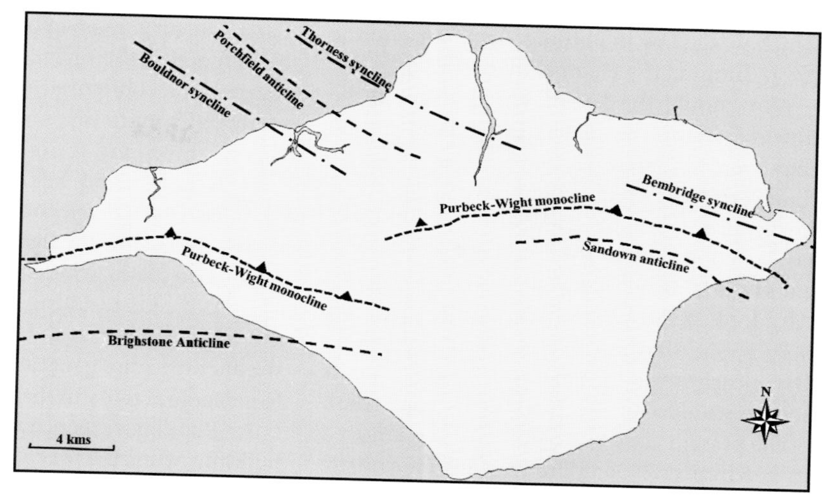

Geological structure of the Isle of Wight.

narrow, reflecting its steep dip on the north side of the monoclinal fold. Marine erosion has created a pointed chalk headland at the Needles, from which extends a series of sea stacks to the west, and on the opposite side of the island is the headland of Culver Cliff.

However, in the middle of the island, southwest of Carisbrooke, the dip of the chalk is lower and the outcrop wider, and here the chalk upland reaches a height of 214m on Brighstone Down. You will also notice that the summits of the chalk hills in both the central and southern outcrops are bevelled, forming a dissected plateau surface. This is thought to be the result of erosion by the Pliocene sea, when these hills became low islands. Subsequently the chalk surface has been deeply dissected by rivers, and since the end of Pleistocene times, the water table has fallen considerably, leaving dry valleys and little surface drainage.

Looking at the geological map, you will see that there is a chalk outlier in the far south of the island where St Boniface Down rises to 266m. The chalk dips gently south and presents a steep, scarped edge to the north, where it overlooks the undulating lowlands of the Lower Greensand, which have been exhumed by the erosion and removal of the chalk that once formed a continuous cover across the southern half of the island. On the south coast near Ventnor, the permeable Chalk and Upper Greensand overlie the impermeable 'blue slipper' or Gault Clay, producing an undercliff. Masses of chalk have slipped down over the wet, lubricated clay to create an undercliff at sea level and produce an uneven bench below the upper cliffs.

Another area of landslipping is the southwest coast from Compton Bay to Blackgang Chine, where the less resistant Wealden sands and clays and the Lower Greensand are being actively undermined by the sea, and frequent slippage occurs on the cliff faces.

The northern part of the Isle of Wight is relatively low-lying and formed of soft Palaeogene sands and marls, except for the resistant Bembridge Limestone that forms the Foreland, the most easterly extremity of the island. This limestone has been used extensively for building in the past, for example in Yarmouth Castle. A glance at the topographical map will show you that the northern shoreline is interrupted by several river estuaries, including the Western Yar, Newtown River, the Medina, Wootton Creek, and the Eastern Yar. These are all classic examples of drowned river mouths, formed when sea levels rose during the Flandrian transgression some 10,000 years ago. At this time the Isle of Wight became separated from the mainland. However, before this happened, these rivers would have extended northwards to join the eastward-flowing 'Solent river', which was fed by the Frome, the Stour and the Avon draining the Hampshire basin.

Why should the Isle of Wight rivers all be flowing north? The answer lies in the fact that they were initiated on the chalk cover (now much eroded), which would have had a gentle northward gradient. Newport and Brading now occupy gaps in the chalk ridge that were cut as the land surface was lowered by erosion. Also look at the Western Yar on the map; it is only about 4km long, rising near Freshwater Bay; clearly it was formerly much longer and its source would have been some distance south of the present coastline before the sea removed its headwaters. Not all drainage is to the north; on the south and west coasts, small streams locally known as chines descend rapidly from the higher ground to the sea, cutting deep ravines in the Greensand.

Plate Tectonics, Folds and Faults

The Earth's surface has evolved over millions of years, and the driving force behind its changing continents is the concept of plate tectonics. It states that the surface of the Earth is composed of large plates that are constantly in motion in respect to one another. Magma continually rises up at the mid-oceanic ridges, cools and solidifies and slowly forces the oceanic plates apart. As these plates move outwards, they eventually collide with adjacent continental plates, where they are forced down in a process known as subduction. This plate movement has been going on throughout the Earth's history, causing oceans and continents to drift across the surface of the globe, so that the present configuration of land and sea is vastly different to that in Precambrian times. When converging tectonic plates collide, one may be subducted below the other, as, for example, where denser oceanic crust is dragged down deep below the lighter continental crust. As the two plates grind together, seismic waves are produced, resulting in earthquakes, and the high pressures and temperatures in the subduction zone generate melts that give rise to volcanic activity in the overlying continental crust. Plate convergence also leads to mountain building, when the rocks may become intensely folded and faulted. This classic scenario can be seen today along the Pacific coast of Peru and Chile, where the Nazca oceanic plate is subducted beneath the continental plate of South America and volcanoes and earthquakes occur within the Andes mountain range.

This process of mountain building, referred to by geologists as orogenesis, has taken place in three major episodes during the last 500 Ma and has had considerable influence on the physical development of Britain. The Caledonian orogeny occurred when the Iapetus Ocean closed as plates converged towards the end of Silurian times, and the subsequent uplift and folding of sediments produced the mountain ranges of Scotland and Wales. Later, at the end of the Carboniferous, the Rheic Ocean closed, initiating the Variscan orogeny that produced intense folding and faulting in southwest England.

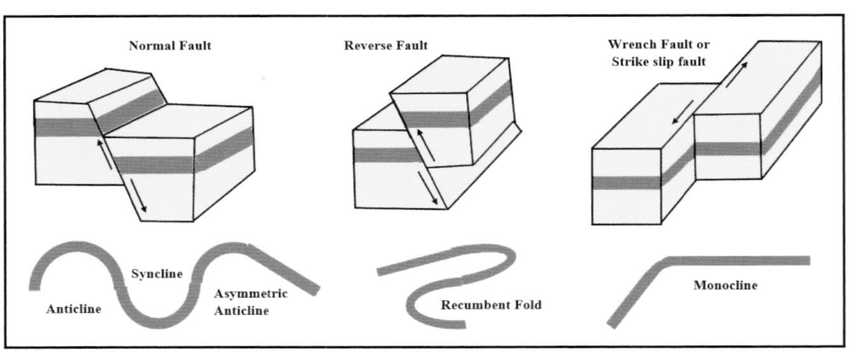

Some of the more common fold and fault structures.

The third great earth building movement, known as the Alpine orogeny, occurred from late Cretaceous to Miocene times as a result of the northward movement of the African and Arabian plates. They were subducted beneath the Eurasian plate as the Tethys Ocean closed, leaving the proto-Mediterranean Sea. This plate convergence squeezed the marine sediments then uplifted and folded them to produce a complex mountain zone including the Pyrenees, the Alps and the Atlas Mountains.

While major uplift took place in these areas, the outer ripples of this orogeny were felt in southern Britain and northeast France. The Weald-Artois anticline and the Purbeck-Wight monocline are impressive testimony to the power of the compressive tectonic forces during the Miocene period, when the Alpine earth movements reached their zenith. The Cretaceous and Palaeogene sediments that occupy the Hampshire Basin (including the Isle of Wight) were all subject to Alpine tectonic disturbance, with pressure being predominantly from the south.

When sediments are deposited, they are usually formed in sub-horizontal layers and, according to the Law of Superposition, the oldest rocks are found at the base of a sequence and successively younger rocks lie above. However, if the layers are later subjected to earth movements, they may be folded and faulted and, in some cases, completely inverted. Anticlines (upfolds) and synclines (downfolds) are the simplest type of fold, but if the lateral pressure is greater on one side of the fold then it will become asymmetric where one limb is steeper than the other. A recumbent fold is one that has been pushed over to such an extent that its lower limb is upside down. Another type of fold is the monocline, where one limb is almost vertical and the other more or less horizontal.

You can simulate the formation of folds simply by moving a carpet on a smooth surface. If you apply equal pressure from both sides you can make symmetrical folds, and by applying more pressure on one side, asymmetrical ones. Compression will not only produce folds, but it will also fracture rock layers to create reverse faults; when rocks are pulled apart under tension, normal faults occur. In both these cases the movement is up or down the dip of the fault plane, but where the movement along the fault plane is horizontal, a wrench or tear fault is produced. This is also called a strike slip fault because the horizontal displacement is parallel to the strike of the fault plane.

A walk along the forshore southeast from Hanover Point. The cliffs are formed by marls of the Wessex Formation, capped by Pleistocene sands and gravels.

CHAPTER 2

Wealden Beds: Land of the Dinosaurs

The oldest beds exposed on the Isle of Wight belong to the Wealden Group (145–126 Ma) of Lower Cretaceous age. They are known as Wealden Beds (in modern terminology, the Wealden Group) because the same strata outcrop in the Weald of Kent and Sussex. They consist mainly of clays and shales with sandstone layers such as are found along the coast of Brighstone Bay in the southwest and behind Sandown Bay on the east side of the island. During Wealden times, rivers draining a land area in the west transported silt, sand and grit to form extensive delta flats with braided distributary channels over much of southern England.

These delta plains were covered with subtropical vegetation, as witnessed by the numerous plant remains found in the sediments. Pieces of wood and pine logs accumulated near the mouth of the delta rivers (distributaries) and are now preserved within the Wealden sediments. Later, as the volume of sand and mud brought down by the rivers decreased due to the erosion of surrounding land masses, large freshwater lagoons emerged, in which finely laminated shales and thin limestones were deposited.

The Wessex Formation

On the Isle of Wight, the Wealden Group is divided into two distinct divisions. The lower beds are known as the Wessex Formation. These consist of red, purple and green marls with some sandstone beds laid down in freshwater conditions on swampy delta tops and wide floodplains crossed by sluggish, meandering rivers. The sandstones were deposited in sinuous river channels while the marls are overbank deposits on the floodplains. The massive Sudmore Point Sandstone that forms the cliffs to the northwest of Chilton Chine is a good example of a channel sandstone; it divides into several sandy horizons separated by marls beyond Sudmore Point, suggesting the possibility that the meandering river became silted up and changed channels over time.

There is much evidence of both flora and fauna within the Wessex beds. The marls contain driftwood, ferns, coniferous tree trunks, and cycad cones, particularly in the plant debris bed that is exposed at beach level near Hanover Point and Brook Chine. There

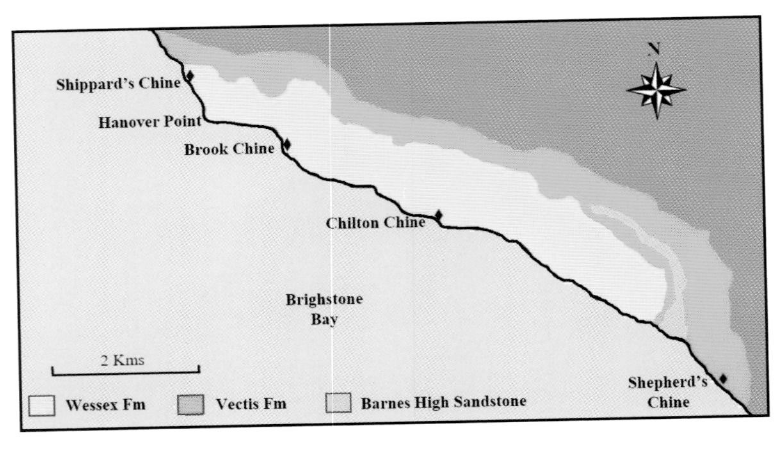

Map of the outcrops of the Wealden Group strata along the southwestern coast of the Isle of Wight.

are also dark lignite layers rich in organic material, which have formed from trees that have fallen into rivers and been washed downstream eventually to be preserved beneath sand and silt. A well-known feature exposed on the foreshore at low spring tides off Hanover Point is the so called 'pine raft'. This is composed of assorted fossilized coniferous logs (gymnosperms) up to 3m in length that are set in the Hanover Point Sandstone. Many of the logs are uncompressed and preserved by carbonate impregnation while others are compressed and carbonized.

The subtropical vegetation growing on the delta flats provided food for large herbivorous dinosaurs such as *Iguanodon*, whose footprints can be seen today on the beach around Hanover Point. The original impressions in the Wealden mud would have filled with sediment and are today preserved as sandstone casts. Dinosaur vertebrae, crocodile and turtle bones, fish teeth and scales also occur in the plant debris beds, but many of the bones are worn and disarticulated. A small freshwater snail (gastropod), known as *Viviparus*, which is found in the lagoonal limestone beds, provides further proof of the environmental conditions at this time. This fossil snail also occurs in the so-called 'Purbeck Marble', which is a beautiful polished limestone much used in church interiors in Dorset and beyond.

The Vectis Formation

The overlying Vectis Formation consists of dark silts and clay with occasional layers of sandstone and shelly limestone that indicate a change to more brackish water conditions in the coastal lagoons as sea level began to rise. The prominent Barnes High Sandstone occurs in the centre of the Vectis sequence; it shows large-scale cross-stratification and is thought to have been deposited in a prograding lagoonal delta. Today it forms a strong cliff top scarp about 500m northwest of Shepherd's Chine. Fossil evidence within the silty sediments includes dinosaur and fish bones, gastropods and bivalves such as oysters and cockle shells. *Filosina* is a common bivalve in the dark mudstone bands of the Vectis Formation.

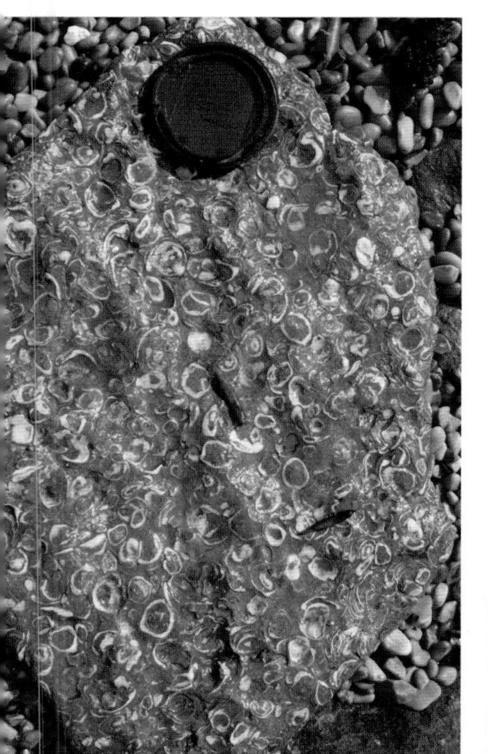

LEFT: **Disarticulated bivalves *Filosina* forming a shell bed in a mudstone band of the Vectis Formation.**

BELOW: **Tridactyl dinosaur footprint, with the cast infilled with sediment, so the block is now upside-down. Footprints and tracks are exposed on the beach at Hanover Point at low tide.**

Itinerary 1: Shippard's Chine to Chilton Chine

Walking distance: 4km
Warning: low tide is required to cross the wave-cut platform at Hanover Point.

You can explore the Wealden Group rocks from several points along the southwest coast. Let us start at the car park above Shippard's Chine [SZ 376842], where slumping in the Wessex marls has removed part of the tarmacked surface. Descend to the beach at low tide and you can stand back and observe that the strata in the cliffs are all dipping northwest and getting younger in that direction. Immediately to the northwest of the chine are the sandstones and clays of the upper part of the Wessex Formation, while to the southeast are the underlying beds extending towards Hanover Point and forming the northern limb of the Brighstone anticline, the axis of which runs through Brook Chine about 500m beyond Hanover Point.

As you approach Hanover Point, look for the plant debris bed that rests on the cross-stratified Hanover sandstone, which is exposed at the base of the cliff. There are fossil trunks and logs of conifers embedded in the sandstone and scattered in all directions forming the 'pine raft'. Much of the material within the plant debris bed is in the form of lignite or carbonized wood.

Wessex marls dipping northwest at Hanover Point.

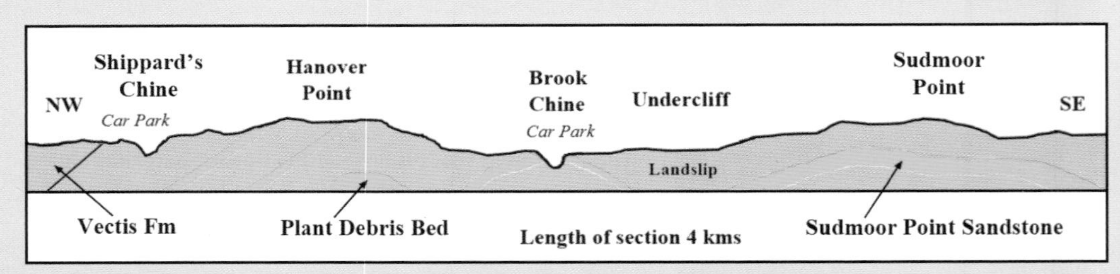

Cliff profile from Shippard's Chine to Sudmore Point exposing Wessex Formation strata.

Sandstones and clays of the Wessex Formation dipping off the Brighstone anticline, near Shippard's Chine.

Brook Undercliff, with rotational slipping in sands and clays of the Wessex Formation where the strata dip gently inland, to the north.

After rounding Hanover Point, make for Brook Chine and the Brook Undercliff. This unstable area is developed in the marls of the Wessex Formation and is subject to landslipping.

The clifftop is 200m inland and is covered by plateau gravels, with its edge scalloped by a series of rotational movements that have produced arcuate depressions and tilted blocks that have slipped down over the mudstones lubricated by spring waters. The broken ground of the undercliff is now a large hummocky zone partly vegetated by gorse and bracken interspersed by areas of bare rubbly sand and clay.

Now move on to Sudmoor Point, where the cliffs are more stable due to the outcropping of a massive resistant sandstone. This Sudmoor Point Sandstone dips southeastwards since we are now on the southern flank of the Brighstone anticline. The sandstone is accessible at beach level about 500m from Chilton Chine. Here the erosional base of the unit can be seen cutting into the underlying red mudstones. The Sudmoor Point Sandstone, up to 6m thick, represents a fining upward sequence from a basal conglomerate through pebbly, coarse, cross-stratified sandstone into fine-grained sandstone with ripple cross-lamination. It represents a point bar deposit formed by lateral accretion in a meandering river channel on a wide floodplain.

On the foreshore, the sandstone shows ripple marks and it is also the location of a dinosaur trackway that is now on display at Dinosaur Isle, the Museum of Isle of Wight Geology in Sandown. Access the main road and car park via Chilton Chine where return transport can be arranged.

View from the northern end of Gore Cliff across Blackgang to Atherfield Point and beyond to Freshwater cliffs.

Dinosaurs in the Wealden Swamps

During Wealden times in the Lower Cretaceous the Isle of Wight enjoyed a subtropical climate, with swamp vegetation flourishing on delta lowlands alongside numerous watercourses. This plant- and water-rich environment of tree ferns, cycads and conifers was home to a variety of creatures, including dinosaurs, pterosaurs, crocodiles, turtles and fish.

At the top of the food chain were the large carnivorous dinosaurs like *Neovenator salerii*. This was a beast that walked on its hind legs with a long tail for balance and, with its three clawed toes and serrated teeth, it would have been a ferocious predator. Some reconstructed skeletons measure 7.5m long by 2.5m high.

Another bipedal flesh-eating dinosaur that has only recently been discovered in the

Itinerary 2: Chilton Chine to Shepherd's Chine

Walking distance: 4km
Warning: the cliffs are unstable and liable to landslides and mudflows; it is advisable to follow this itinerary on a falling tide.
This next section of the 'Wealden coast' begins at the car park at Chilton Chine [SZ 410822], from where there is easy access to the beach. As you walk southeastwards, you will notice that the cliffs are composed mostly of red mudstones interbedded with some thin sandstone units – all part of the Wessex Formation. The Brighstone Sandstone appears high in the cliffs and dips down to beach level about 500m beyond Chilton Chine.

The mudstones above the sandstone appear to be considerably disturbed, and this is an example of what geologists call soft sediment deformation. During deposition, when the sediment is still liquified, slumping may occur. Gravity-induced structures may also develop as convolute bedding in which multiple folds are produced.

Cross the outlet to Grange Chine and look in the cliffs for the Grange Chine Black Band, a conspicuous grey clay layer containing carbonized plant debris. It is thought that this and other similar bands were formed by sheet flooding and debris flows that swept fallen vegetation downstream into coastal swamps and lagoons. Continue along the beach towards Barnes Chine, pausing at Ship Ledge (formed by a fine-grained white sandstone) to view the cliffs at Barnes High, which rise to over 50m above sea level. Here the Barnes High Sandstone forms a spectacular cliff crest escarpment, below which are the basal mudstones of the Vectis Formation.

Our next stop will be about 100m before Cowleaze Chine, where the contact between the Wessex and Vectis formations is at beach level. You cannot miss it because there is a distinct change in colour from the underlying reddish Wessex marls to the overlying grey Vectis mudstones.

Continue along the shore to Shepherd's Chine, where the Barnes High Sandstone reaches the beach. Look carefully and you will see that this 6m-thick sandstone unit shows a coarsening upwards sequence: starting with fine sand with wavy, lenticular bedding, the grain size increases towards the top, where cross-stratification and ripple bedding become increasingly larger in scale. These sedimentary structures suggest that the sandstone was laid down in a coastal environment possibly as a barrier bar or on tidal flats. You can gain access to the A3055, where there is roadside parking, by taking the footpath up through Shepherd's Chine.

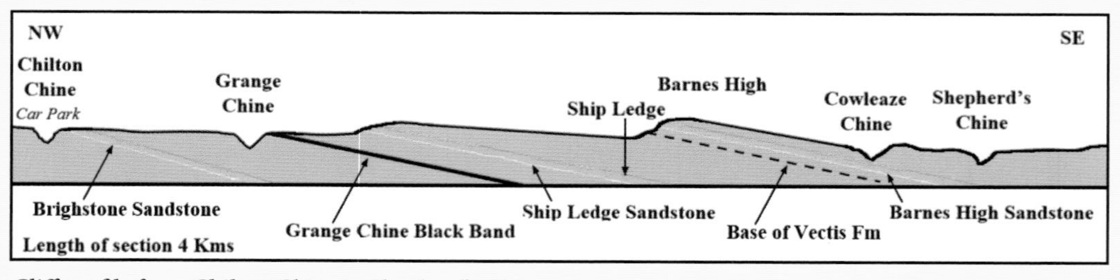

Cliff profile from Chilton Chine to Shepherd's Chine.

Wealden Beds is *Eotyrranus lengi*. This was one of the forerunners of the well-known *Tyrannosaurus rex* and it is unusual in having long hind legs for running and forearms with which to grasp its prey. It was probably one of the fastest predators on the island at the time, terrorizing the local herbivores!

Iguanodon bernissartensis is perhaps the most famous dinosaur found on the Isle of Wight, although its specific name comes from a small Belgian town where the first complete skeleton was discovered in 1878. This herbivorous dinosaur stood up to 4m high and weighed 4–5 tonnes. For over a century it was thought to be bipedal, standing on its hind legs to reach the hanging branches of tall trees. However, more recent research with the aid of computer graphics and biodynamics has shown that it would have been able to walk on all fours, with its spine almost horizontal and tail held out for balance. Even so, the tridactyl footprints of hind feet found on the foreshore from Compton Bay to Atherfield Point suggest that these large herbivores spent much time in an upright posture reaching for foliage.

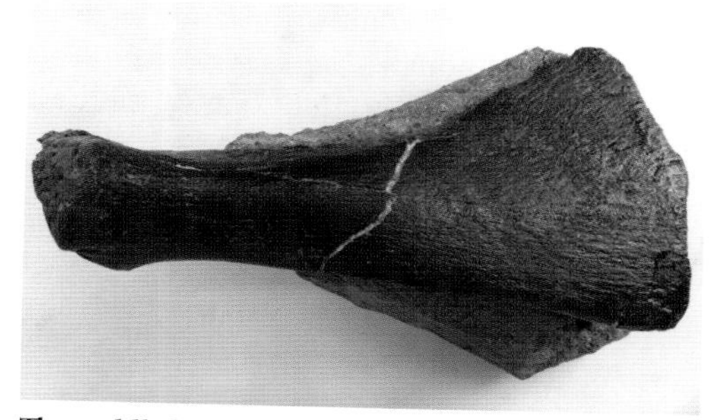

The paddle bone of a plesiosaur, a large aquatic reptile, found in the Wealden beds at Yaverland by Alex Peaker of the Dinosaur Isle Museum.

In 1917, a lighter, more graceful specimen was found in the Atherfield Clay and later named *Iguanodon atherfieldensis* but the latest research has shown that this skeleton was significantly different from *bernissartensis* and it is now placed in a new genus and renamed *Mantellisaurus atherfieldensis*.

In 1992, an almost complete skeleton of a giant Sauropod was unearthed in the Wealden rocks of the Isle of Wight. This creature appears to have walked on all fours and it had a long neck, enabling it to reach overhanging

The skeleton of *Iguanodon bernissartensis* displayed within the Dinosaur Isle Museum at Sandown. This magnificent reconstruction incorporates a large number of bones that were found in Lower Cretaceous sediments exposed along the coast and now assembled on a wall of imitation rock.

vegetation. There is some uncertainty as to its exact name, although *Pelorosaurus* has been suggested; but it is certainly one of the best Sauropod specimens ever found.

If you wish to see dinosaur specimens and other fossils from the Isle of Wight you should visit the superb purpose-built interactive museum at Dinosaur Isle in Sandown.

How Old are the Rocks?

We read that rocks are hundreds of millions of years old, but how do we know this? In 1650, Bishop Ussher, used biblical chronology, to calculate that Earth was created in 4404BC. But his views fell into disrepute as the concept of uniformitarianism developed. This suggested that the Earth's surface had changed extremely slowly and that the geological processes at work today have been happening in the past.

Indeed, the great 18th-century Scottish philosopher and scientist James Hutton believed that 'the present is the key to the past'. When Hutton discovered his famous unconformity at Siccar Point, Berwickshire, he was convinced that it represented an enormous time gap of millions of years. In fact, the age gap represented by the Siccar unconformity is about 65 million years, from Lower Silurian greywacke (435 Ma) to the overlying Upper Old Red Sandstone. In his book *Theory of the Earth*, Hutton was so convinced of the Earth's great age that he concluded with the words, 'No vestige of a beginning; no prospect of an end'.

Throughout the nineteenth century, geologists developed a stratigraphical sequence that provided a relative timescale. William Smith surveyed canal routes in Somerset and showed that each set of rocks contained a distinct set of fossils. This led to the establishment of a geological chronology divided by palaeontologists into eras and periods, with the oldest rocks at the base and the newest at the top, according to the Law of Superposition. Major boundaries in the Earth's timescale coincided with major extinction events, when certain lifeforms were wiped out. This division of time according to relative position in the rock record is known as the chronostratigraphic timescale.

However, it was not until the early twentieth century that the science of geochronology was developed, allowing geologists to calculate the absolute age of rocks. Professor Arthur Holmes was a pioneer in geochronology and he was the first to perform an accurate uranium-lead radiometric dating. When a radioactive isotope of uranium decays (or is converted) into an isotope of lead, the rate of decay is known; it is a constant that can be calculated. Half of the uranium isotope will decay in a given time period, which is known as its half-life. Since the rate of decay for uranium is very slow, its half-life is around 4,000 million years. The parent isotope must have originated when the uranium first crystallized. The radiometric method compares the amount of a naturally occurring isotope remaining within a rock sample to the amount of its decay products, which form at a constant known rate and can be measured; thus the period of time during which the mineral has been decaying can be calculated. This is the radiometric date of the rock, which, for example, could be 400 Ma.

Other elements apart from uranium can be used in radiometric dating; they include potassium, which decays to argon, and carbon 14, which decays to nitrogen. The latter only has a short half-life and so it is used to date more recent organic material, such as plants and fossils younger than 60,000 years old.

Due to the development of sophisticated radiometric techniques yielding absolute dates for the age of the rocks, geologists can now assign more accurate dates to the periods and formations within the stratigraphical column. It can even be stated with confidence that the age of the earth is at least 4.5 billion years, a far cry from the days of Bishop Ussher! Despite these advances in absolute dating, to the geologist in the field it is often more important to understand the relative ages of rocks, their relationship to each other and the palaeoenvironments in which they were formed.

CHAPTER 3

Tropical Seas of the Lower Greensand

Greensand is a sandstone containing the green mineral glauconite, a silicate of iron, aluminium and potassium that forms under shallow marine conditions. It was first recognized and named in the eighteenth century before it was realized that many outcrops of greensand are coloured red, brown or orange because they have been oxidized when exposed on the surface. It is also worthy of note that there are two greensands within the Cretaceous succession: the Upper Greensand above the Gault Clay and the Lower Greensand below. The Lower Greensand underlies a major part of the southern half of the Isle of Wight and is well exposed on the coasts of Sandown Bay and Chale Bay.

At the beginning of Lower Greensand times (126–113 Ma), the Vectis delta flats would have been gradually subsiding and, as sea level began to rise, the sea flooded over the swampy lowlands. We have first-hand evidence of this marine transgression from the Perna Bed, composed of a thin band of sandy clay overlain by a calcareous sandstone that rests on the erosion surface of the Vectis Formation. The sandstone is full of marine fossils, including oysters and the large bivalve *Perna mulleti*, after which the bed is named. The Perna Bed can be seen at the base of the Lower Greensand

View southeast across Chale Bay. The Ferruginous Sands weather to create the orange-red beach sands.

at Atherfield Point and also in the Yaverland cliffs. When the sea began to deepen, the Perna Bed was overlain by muddy sediments, which we see today as the Atherfield Clay.

There are several fossiliferous beds within the clay, including the Lower and Upper Lobster beds separated by the Crackers. The latter are two lines of large, rounded calcareous concretions (also called doggers) that form a prominent feature in the cliff near the Atherfield coastguard station. When the concretions fall from the cliffs they are often broken open by the waves to reveal fossils within the softer cores. Well-preserved ammonites, gastropods, bivalves and echinoids may be seen inside the concretions. In addition to these delights, the Lower Lobster Bed is famed for its beautiful specimens of the fossil lobster *Meyeria vectensis* and the fossil prawn *Meyeria magna*.

The Atherfield Clay is overlain by the Ferruginous Sands Formation, a thick sequence of alternating clays and glauconitic sands occurring in rhythmic coarsening upward units, reflecting fluctuating sea levels in relatively shallow waters. These ferruginous beds (Latin *ferrum*: iron) have weathered to red- and brown-coloured rocks, as can be seen at Red Cliff northeast of Sandown and on the Chale-Blackgang coast. The strata can be divided into the following stratigraphical sub-units that are distinguished along the Chale Bay coast. These divisions were originally set out by W. H. Fitton in 1847 and can still be identified today. Some of these fossiliferous beds are described below:

Lower Gryphaea Beds are named after the large oyster *Aetostreon* (formerly known as *Exogyra*), which was once mistaken for *Gryphaea*. The oysters tend to be concentrated at particular horizons (firmground) where there are also nests of small terebratulid brachiopods (for example *Sellithyris*) in life position.

Scaphites Beds are well known for their bands of nodules containing the large ammonite *Australiceras*, which belongs to the same family as the loosely coiled Scaphites.

Lower Crioceras Beds contain many concretions, inside which are often found bivalves and ammonites, including the large crioceratid *Tropaeum*, with its distinctive open spiral form.

Upper Crioceras Beds are rich in ammonites and bivalves and also contain plant remains such as fronds of the fossil fern *Weichselia*.

Walpen and Ladder Chine Sands are distinguished by a line of concretions with phosphatic cores full of calcite-filled ammonites, bivalves and brachiopods.

The succeeding Sandrock Formation is paler in colour and less iron-rich than the Ferruginous Sands and forms the cliffs near Rocken End and at Knock Cliff near Shanklin. The sands and muds appear to have been deposited under estuarine conditions. The trace fossils (infilled burrows) suggest a littoral situation and cross-stratified sands are consistent with high-energy shallow water. The overlying Carstone Formation consists of coarse-grained ferruginous sandstone with seams of concretionary ironstone. The BGS have recently replaced the traditional term 'Carstone' with the name of the type locality – Monk's Bay Formation.

Top of Ferruginous Sands Formation
Ferruginous Bands (Blackgang Chine)
Walden Undercliff Sands
Laminated Clays and Sands
Cliff End Sand
Upper Gryphaea Beds
Walpen and Ladder Chine Sands
Upper Crioceras Beds
Walpen Clay and Sand
Lower Crioceras Beds
Scaphites Beds
Lower Gryphaea Beds
Base of Ferruginous Sands Formation

The sequence of strata that constitute the Ferruginous Sands, as recognized by W. H. Fitton.

Pioneering Geologists on the Island

Robert Hooke (1635–1703) One of the earliest natural scientists associated with the Isle of Wight was born at Freshwater. He became Curator of Experiments at the Royal Society in London, where he made various scientific advances, including improvements to pendulum clocks and the development of a compound microscope. He studied plant cells in fossilized wood and the detailed structure of insects using his new microscope. However, his best-known contribution was the publication of *Micrographia*, a seventeenth-century bestseller revealing the latest in scientific progress.

Gideon Mantell (1790–1852) Another early pioneer, Mantell worked as a country doctor in Lewes, Sussex. He was inspired by Mary Anning's sensational discovery of *Ichthyosaur* remains at Lyme Regis and spent much of his spare time investigating the Cretaceous strata around the Weald, where he eventually discovered fossil teeth that were initially thought to be from fish. After much dispute in discussion with other scientists including William Buckland, the Oxford geologist, Mantell was convinced that the teeth and bones that he unearthed belonged to a huge reptilian creature with teeth resembling those of iguana, the South American lizard.

Gideon Mantell (1790–1852), obstetrician and palaeontologist, who made the first scientific study of dinosaurs.

Finally, after corresponding with the palaeontologist W.D. Conybeare, the latter suggested the name *Iguanodon* and one of the most famous dinosaurs in the world entered scientific history. As more fossil remains were discovered, Mantell was able to demonstrate that *Iguanodon*'s forelimbs were much shorter than its hind legs, suggesting a bipedal stance enabling it to graze on higher vegetation. It is interesting to note that it was several years after Mantell's death that Sir Richard Owen, a specialist in vertebrate palaeontology, first coined the term 'Dinosauria' (terrible lizard) based on the three genera of *Megalosaurus*, *Iguanodon* and *Hylaeosaurus* that had been uncovered in southern England.

Mantell wrote several scientific papers on the dinosaurs of the Isle of Wight and, in 1847, during the latter part of his life, he published a geological guide entitled 'Geological excursions round the Isle of Wight and along the adjacent coast of Dorsetshire'. He also had several fossils named after him including *Mantellisaurus atherfieldensis* (formerly *Iguanodon atherfieldensis*); *Lepidotes mantelli*, a ray finned fish from the Lower Cretaceous; and the ammonite *Mantelliceras mantelli* from the Lower Chalk.

William H. Fitton (1780–1861) An Irish medical practitioner, Fitton only became a geologist in later life. He was president of the Geological Society of London and in 1836 presented an important monograph 'Observations on some of the strata between the Chalk and the Oxford Oolite in the southeast of England'. William Smith had already published his geological map of England and

continued on p.30

Itinerary 3: Shepherd's Chine to Rocken End

Walking distance: 12km return
Warning: this is a long trek and should only be started on a falling tide and in calm weather. You can of course, opt to do only part of the itinerary. The cliffs can be dangerous due to rock falls and mudslides. Hard hats are recommended below cliffs, especially after heavy rain.

There are only a few places where you can access the Lower Greensand coast in the southwest. Shepherd's Chine has roadside parking [SZ 451799] and is the best place to start. Descend to the beach through the chine and walk towards Atherfield Point, passing the cliffs cut into the Vectis mudstones that dip gently southeast.

As you approach Atherfield Point, you will see that there is a calcareous sandstone reef marking the outcrop of the Perna Bed that forms the base of the Lower Greensand. The boulders on the foreshore provide a rich source of fossils including bivalves, gastropods and brachiopods. Trace fossils are numerous, including the horizontal branching burrows known as *Thalassinoides*. These burrows were excavated by crustaceans (crabs, shrimps and so on) and are indicative of shallow water environments.

The Atherfield Clay Formation overlies the Perna Bed and forms the steep crumbling headland, with scree cones of weathered talus accumulating at the base of the cliffs. The Lower Lobster Bed that succeeds the Atherfield Clay is marked by a resistant sandstone ledge; then we approach the outcrop of the Crackers Beds, yellow-brown sandy clays with two prominent layers of rounded concretions. The Crackers beds reach the beach approximately 500m southeast of the coastguard station and they provide an easily recognizable marker within the Lower Greensand. If you feel strong enough to carry a sledgehammer, it is possible to crack open some of the smaller nodules to reveal beautifully preserved ammonites, bivalves and gastropods within the soft cores.

Trace fossils in bioturbated sandstone from the Ferruginous Sands Formation showing three-dimensional burrows infilled with sediment.

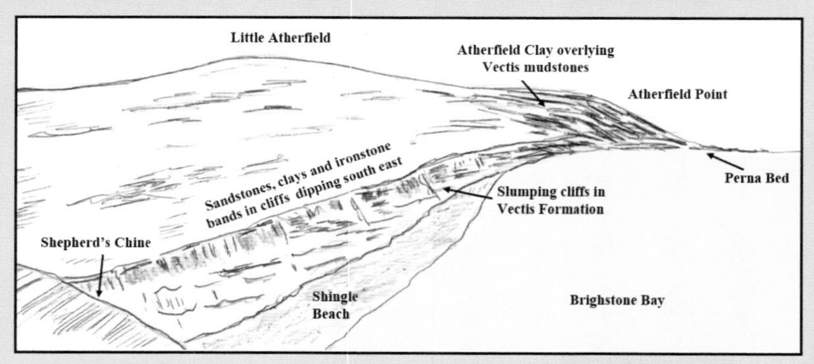

LEFT: **Field sketch looking towards Atherfield Point from above Shepherd's Chine, with slumping cliffs of Wealden Beds (Vectis Formation) in the foreground.**

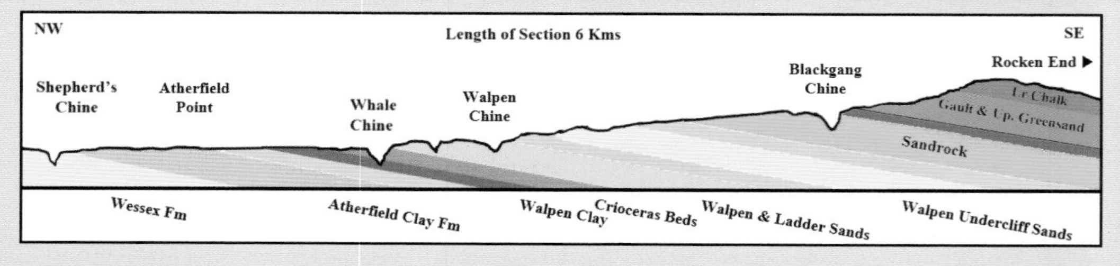

BELOW: **Cliff profile from Shepherd's Chine to Rocken End. Strata from the Atherfield Clay to the base of the Sandrock are in the Ferruginous Sands.**

RIGHT: **Calcareous sandstone concretions within the Ferruginous Sands Formation in Chale Bay. The concretions lie along a bedding plane dipping towards the beach. They form as mineral segregations during diagenesis; here they are more strongly cemented on the outside than in the core.**

BELOW: **Whale Chine, with almost horizontal beds of Ferruginous Sands well exposed on its steep sides.**

As we get about 300m from Whale Chine, the Ferruginous Sands Formation appears at beach level and there are striking exposures of the red, iron-rich Crioceras Beds in the walls of the upper part of Whale Chine.

The lower section of the chine is best seen from beach level where a waterfall flows over a steep cliff formed by a hard sandstone band at the top of the dark Walpen Clays. You should be aware that although there is a car park above Whale Chine [SZ 470784] there is no access to the beach since wooden steps have been destroyed by cliff recession, and access to the public is now prohibited.

BELOW: **The cliffs around Whale Chine looking towards Gore Cliff and Rocken End. A recent landslip has brought blocks of Ferruginous Sandstone down onto the beach.**

A bed of iron-stained oysters *Aetostreon* (syn. *Exogyra*) from the Ferruginous Sands. These are sometimes referred to as 'snuff boxes' similar to those found at Burton Cliff, Dorset.

Continue towards Ladder Chine and Walpen Chine, which are notches cut into the cliff line where the Walpen and Ladder Sands dip gently southeast in the cliff face. These uniform grey sands commonly yield the large oyster *Aetostreon* resting on firmground horizons.

ABOVE: **Whale Chine, a fine valley incised as a result of stream downcutting, induced by cliff recession. A waterfall flows over a hard clay band in the Walpen Clays that form the lower part of the cliff. The wider valley above is cut through the red Crioceras Beds.**

View southeast of Whale Chine, showing Ladder Chine and Walpen Chine notched into the cliff line and looking towards Rocken End. The Ferruginous Sand Formation includes the dark Walpen Clays that outcrop at beach level and the light-coloured Walpen and Ladder Sands that form the upper cliff.

As you approach Blackgang, you will see the Sandrock dominating the upper slopes that overlie the Walpen Undercliff. This cliff overlooks an undercliff terrace over 100m wide that is covered with a mudslide of slumped debris and sandy outwash. A spring line occurs at the back of the terrace, fed by groundwater percolating through the Walpen Undercliff Sands. This seepage is responsible for the undermining, collapse and recession of the upper cliffs and the consequent disappearance of Blackgang Chine. It is now only possible to examine the lower part of the sequence since the cliffs here have been subject to numerous landslips and mudslides that obscure the strata.

ABOVE: **A fallen block of weathered sandstone from the Ferruginous Sands containing numerous oysters infilled with sand.**

LEFT: **Cliff section through the former Blackgang Chine showing mudslides fed by springs at the base of the Walpen Undercliff.**

BELOW: **Mudslides at Blackgang Chine. The prominent Sandrock cliff can be seen above the Undercliff terrace.**

Itinerary 4: Yaverland to Culver Cliff

Walking distance: 4km

This itinerary begins at Yaverland car park [SZ 611849], opposite Sandown Zoo. The Lower Cretaceous sequence exposed along this stretch of coast lies on the northern flank of the Sandown anticline and the strata dips NNE increasing from about 10 degrees at Yaverland to almost vertical at Culver Cliff.

Walk down to the beach, where the Wessex Formation is seen in the low crumbling cliffs to the east. The sequence includes red mottled mudstones interbedded with sandstones representing floodplain deposits. Some plant debris beds are scattered through the succession; they contain carbonized fragments of coniferous wood, crushed cones and sometimes the remains of tree ferns.

About 500m from the car park, there is a distinct change of colour in the cliffs; this is where the boundary between the red Wessex beds and the grey Vectis mudstone occurs at beach level. Just beyond this junction you should be able to locate the Barnes High Sandstone, which forms a low cliff with conspicuous beds of cross-stratified yellow sands. The upper part of the Vectis Formation is much obscured by the Yaverland landslide that has developed in the Atherfield Clay but it is possible to locate the Perna Bed at the base of the clay. This forms a ridge of sandstone on the foreshore about 1km east of the car park. The upper part of the Perna Bed is richly fossiliferous, including large bivalves such as *Mulletia* (formerly *Perna*) and the coral *Holocystis*.

The next stop further along the beach is at Red Cliff, where the Ferruginous Sands form a headland of resistant strata.

The red-brown colour of the rocks reflects the oxidation of the iron compounds glauconite and limonite due to weathering. The paler Sandrock beds overlie the Ferruginous Sands on the northeast side of Red Cliff, and the Lower Greensand succession is then completed by the Carstone, an iron-rich sandstone.

Between here and Culver Cliff you pass the slumped mass of Gault Clay, followed by the Upper Greensand, before reaching the chalk bastion of Culver Cliff itself (*see* Chapter 4).

ABOVE: **Iron-rich sandstones dip steeply north at Red Cliff on the northern flank of the Sandown anticline.**

BELOW: **Slumped Atherfield Clay underlies Red Cliff. The concrete blocks and brickwork on the beach are due to cliff erosion.**

ABOVE: **Carstone is a red-brown ferruginous sandstone at the top of the Lower Greensand. The contorted bands of dark ironstone are more resistant to erosion than the paler sandstone.**

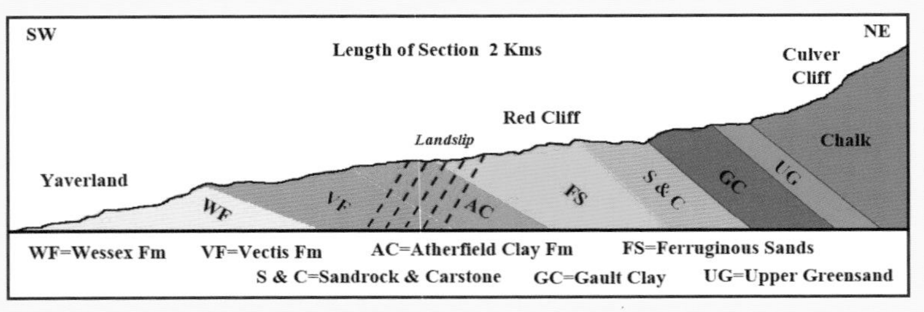

LEFT: **Cliff profile from Yaverland to Culver Cliff.**

Itinerary 5: Sandown Pier to Monk's Bay via Knock Cliff

Walking distance: 6km

A suitable place to start this walk is Sandown pier [SZ597840]. The axis of the Sandown anticline, orientated roughly east–west, crosses the coast near Sandown Zoo, where the Wealden Beds form the core of the anticline. As we walk southwards, the overlying beds of Lower Greensand dip gently to the south because we are on the southern flank of the anticline.

Sandown is built on a low plateau of Ferruginous Sands (within the Lower Greensand), which form steep cliffs along the shoreline. The Undercliff Walk runs in front of the cliffs of muddy sandstone with lines of ferruginous concretions that are thought to be equivalent to those exposed in Whale Chine (*see* Itinerary 3). These are succeeded by yellow weathered sands, some of which show large-scale cross-stratification. At Little Stairs [SZ 589824], there is a fault complex that downthrows to the south and brings down the Walpen Undercliff Sands to beach level. These sands are near the top of the Ferruginous Sands Formation. Here the strata are almost horizontal, and beyond Small Hope Chine the low cliffs are largely covered by buildings.

Continue along the Esplanade that ends near Shanklin Chine, a steep wooded valley cut into the Walpen Undercliff Sands, which display large-scale cross-stratification. Knock Cliff [SZ 585811] is about 750m to the south, opposite Horse Ledge. The latter is formed of the hard *Exogyra* sandstone that extends seawards as a wide intertidal platform dipping gently southwards. The top of this bed is quite fossiliferous, containing the brachiopods *Rhynchonella* and *Vectella*, the oyster *Aetostreon* (formerly *Exogyra*) and the echinoid *Toxaster*. The brachiopods occur in groups that appear to be in life position.

The lower part of Knock Cliff is formed of the uppermost beds of the Ferruginous Sands (equivalent of the Ferruginous Bands of Blackgang Chine), containing dark sands with lines of phosphatic concretions often enclosing bivalve and gastropod internal and external moulds. Although the concretions can only be examined in situ at beach level near Yellow Ledge, many loose examples can be picked up off the foreshore. There is a grass-covered bench feature about halfway up Knock Cliff that is formed by an impermeable bed of sandy clay at the base of the overlying grey Sandrock Beds, from which there is a constant seepage of groundwater.

At low tide you can walk around Yellow Ledge and into Luccombe Bay, where the Sandrock Beds can be seen at beach level. There is an interesting pebbly sandstone bed about 3m above the beach that contains phosphatic nodules and fossil wood

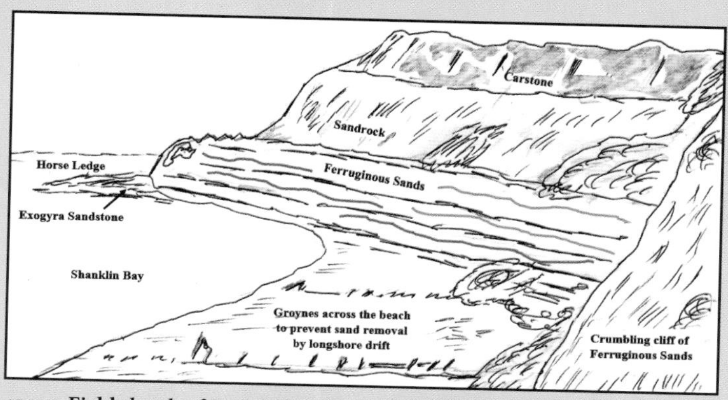

ABOVE: **Field sketch of Knock Cliff viewed from Appley Steps near Shanklin, showing Ferruginous Sands overlain by Sandrock and Carstone Beds.**

BELOW: **Section from Shanklin Chine to Luccombe Chine.**

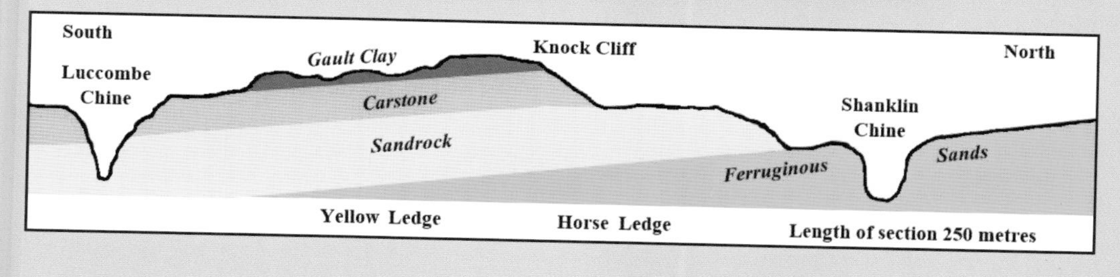

fragments, including some cones. The nodules also yield a rich flora of cycads and conifers. The red-brown Carstone with a capping of Gault Clay outcrops at the top of the cliffs on either side of the wooded Luccombe Chine.

You can walk up the chine into Luccombe village, where there are tea rooms, then return to Shanklin via the coast path. Alternatively, it is possible at low tide to scramble across the foreshore boulders to reach Bordwood Ledge, but beware of being cut off by the incoming tide. The Sandrock Beds are covered by slipped masses of Gault Clay at the foot of the cliffs behind the wave-cut platform, where there are boulders of both Sandrock and Carstone derived from higher up the cliffs.

From here south to Dunnose, extensive landslips have occurred where the Lower Chalk and Upper Greensand have slipped over the impermeable Gault Clay as groundwater has lubricated the rotational slip planes. The best way to view the landslips from above is to return to Luccombe Chine and take the coast path south through the wooded hummocky undercliff of descending terraces to Monk's Bay, where the Carstone and underlying light-coloured cross-bedded Sandrock Beds appear again in the cliffs. This is the type area for the Sandrock that, for obvious reasons, has recently been renamed as the Monk's Bay Sandstone. It is a short walk from Monk's Bay up into Bonchurch, where refreshments and transport are usually available.

Knock Cliff, extending out to Horse Ledge. Ferruginous Sands in the lower cliffs are overlain by buff Sandrock.

continued from p.23

Wales (1815) but Fitton provided much more detail on Upper Jurassic and Lower Cretaceous rocks. By 1847, he had explored much of the Isle of Wight and published a paper called 'A stratigraphical account from Atherfield to Rocken End on the southwest coast of the Isle of Wight' in the quarterly journal of the Geological Society.

His work on the Ferruginous Sands of the Lower Greensand resulted in the establishment of twelve subdivisions, many of which are still recognized today. The lowest subdivision, known as the Crackers Bed (now included in the Atherfield Clay Formation), was named by Fitton because of the explosive noise caused by the waves breaking in air pockets within cavities below a line of concretions.

Edward Forbes (1815–54) Forbes was a palaeontologist for the Geological Survey and professor of natural history at the University of Edinburgh. He worked on the Palaeogene rocks of the north coast of the Isle of Wight and in 1853 published a paper 'On the Tertiary fluvio-marine formation of the Isle of Wight'.

There are other famous nineteenth-century scientists that have links with the Isle of Wight:

Charles Darwin, who revolutionized so much scientific thought with his theory of evolution and natural selection, began writing an outline of his *On the Origin of Species* while staying at a hotel in Sandown in 1858.

Sir Charles Lyell, the great Scottish geologist, refers to the physical structure of the island in his seminal work *Principles of Geology*, published in 1830.

Sir Joseph Prestwich was an authority on the Palaeogene strata of both southern England and northern France. In 1846 he published a study 'On the Tertiary or Supercretaceous formations of the Isle of Wight as exhibited in the sections at Alum Bay and Whitecliff Bay'.

What do Fossils tell us?

In North Yorkshire, fossil ammonites are abundant in the Lower Jurassic rocks; the legend goes that the Saxon abbess St Hilda of Whitby (AD614–680) turned all the local sea serpents to stone – and that is where they come from! In Medieval times fossil sea urchins were considered good luck charms, and so they were placed by the doors of houses and churches. Echinoids were also called 'thunderstones' and were thought to have descended from heaven during thunderstorms.

In the seventeenth century, some natural philosophers referred to fossils as *lapides figurati* or figured stones, although Robert Hooke (1635–1703) clearly recognized them to be the petrified remains of ancient creatures, some of which no longer existed. William Smith (1769–1839), the canal engineer often called 'the father of English geology', demonstrated that strata of different ages contained different assemblages of fossils. Furthermore, he showed that rocks could be correlated over considerable distances according to the fossils they contained.

Karl Linnaeus (1707–78) had already introduced a system of hierarchical classification of plants and backboned animals but he gave scant recognition to invertebrate creatures. Darwin introduced the concept of natural selection and evolution using fossil evidence that was available when he published his *Origin of Species* in 1859. By the middle of the nineteenth century, palaeontology, or the study of prehistoric life, had become established as a scientific discipline and public interest in the subject was stimulated by the opening of the Natural History Museum in Kensington in 1881 (the brainchild of Sir Richard Owen, the illustrious Victorian palaeontologist).

Marine Invertebrates

Although dinosaurs have long been popular in the public imagination, modern palaeontology is largely focused on the vast range of invertebrate fossils. So what do the remains of long dead creatures tell us about the past and what use are they to the geologist?

Fossils can be used to reconstruct the palaeoenvironment. 'The present is the key to the past' was the dictum of James Hutton, the famous eighteenth-century Scottish geologist. Many fossils have living relatives, which provide clues to the mode of life in the past. Some bivalves, for example, may burrow into soft mud on the sea floor; gastropods may graze on hard rock surfaces, brachiopods attach themselves to rocks. Marine invertebrates, such as corals, brachiopods, ammonites and echinoids, mostly required water of normal salinity. However, some adapted to life in estuaries and lagoons, such as gastropods and bivalves. Reef-forming corals are indicators of warm water 25°–29°C

Trilobites were common in the Lower Palaeozoic era. Many of these were adapted to crawling on the muddy seafloor and developed compound eyes.

Micraster, **an echinoid known as the 'heart urchin' found in the Chalk and often infilled with flint.**

up to 90m deep where the waters are well oxygenated and sediment-free. Particular bivalves may be restricted to warm tropical waters or alternatively they may be adapted to life in arctic waters. Some trilobites had complex eyes to see in the murky depths of the sea, some were blind and others adapted for swimming.

Index fossils evolve rapidly, having a short vertical time range, and they have a wide geographical distribution, enabling correlation of strata over a wide area. Relatively abundant and well-preserved specimens usually prove to be good time indicators that can identify certain stratigraphical zones and distinguish one zone from another. *Micraster*, the heart urchin, is a classic example of an evolving species from the Upper Cretaceous. Its morphology changes through the Upper Chalk and so it is used to define several zones including *M. cortestudinarium* and *M. coranguinum*.

Ammonites are also useful index fossils and they are used extensively to subdivide Mesozoic strata since their morphology changes considerably over time. Some are tightly coiled like *Dactylioceras*, others have strong ribs (*Asteroceras*) or a prominent keel (*Hildoceras*), while towards the end of the Cretaceous, uncoiled ammonites like *Scaphites* had evolved.

But let us not forget that fossilization potential is important when selecting useful fossils. Why are some fossils preserved when most are destroyed? Some creatures suffered catastrophic burial beneath sudden mudslides and are preserved in life position, others were transported by currents after death and often fragmented. Soft parts of animals usually decay rapidly when they die and even fossil shells may dissolve, leaving a space that fills with sediment to form an internal cast. Mineral-rich waters may seep into a shell and infill the space with crystalline material. The chambers within ammonites may be infilled with calcite, echinoids in chalk may be replaced by flint and shells may be pyritized (coated in iron pyrites, FeS_2) by burial under anoxic conditions.

An assemblage of ammonites that were preserved on the sea floor by rapid deposition of sediment.

CHAPTER 4

The Chalk Seas

Towards the end of the Lower Cretaceous, there was marine transgression across much of southern England and, as sea level started to rise, the Gault Clay was deposited. There is little evidence of current movement, hence it appears that the silty clays were formed as slow sedimentation took place in quiet conditions. Ammonites, gastropods and bivalves provide examples of marine fauna. However, the Gault is only about 30m thick on the Isle of Wight and so it produces narrow outcrops below the Upper Greensand and Chalk escarpments. It is notable for its impermeability, which causes the overlying rocks to slide over its wet surface, resulting in landslides particularly around the south coast. Gore Cliff above Rocken End suffered a major rock fall and landslide in 1928, destroying the old coast road.

The Gault Clay passes up into the Upper Greensand (113–100 Ma), which begins with a silty sand unit known as the 'passage beds' and then becomes essentially a glauconitic sandstone with frequent bands of calcareous concretions and some layers of hard phosphatic nodules. The upper part of the formation is referred to as the Chert Beds, where the greensand is characterized by both concretions and regular layers of chert. The chert appears to have been formed as segregations of siliceous solutions accumulated in certain horizons. The silica was probably sourced from sponges that were abundant in the marine environment of the greensand but there is still some controversy over the exact origin of the chert.

The Upper Greensand forms a prominent bench feature below the chalk escarpment as, for example, on the western side of St Catherine's Hill. The grey-green sandstone also provides a useful building stone widely used in local churches, cottages and walls.

After the Upper Greensand was laid down, the seas started to deepen as the widespread Cenomanian marine transgression began. This event initiated the deposition of calcareous sediment across much of Britain and northern Europe. Deposition continued for about 35 million years to the end of the Cretaceous period (66 Ma), and hundreds of metres of chalk accumulated as sea level increased; perhaps to 300m above today's level. The great thickness of chalk strata implies that as the sea deepened, the sea floor was subsiding under the weight of sediment.

Initially, calcareous mud derived from the surrounding land produced the Glauconitic Marl, the basal division of the Lower Chalk; the bulk of this subgroup consists of numerous rhythmic alternate metre scale units of marl grading up into chalk and is known as the Chalk Marl. The ammonite *Schloenbachia* and the sponge *Exanthesis* are examples of fossils found within the marls. The succeeding pale grey Zig Zag

Irregular layers of dark chert set in glauconitic sandstone in the Upper Greensand of Gore Cliff.

Chalk is topped by the distinctive Plenus Marls that mark the upper limit of the Lower Chalk.

As the sea deepened and the land was far away, the sediments became almost pure calcium carbonate, made up of the skeletal remains of microscopic planktonic algae called coccoliths and shell fragments. Consequently, the White Chalk subgroup (formerly Upper and Middle Chalk) is characterized by thick beds of pure white limestone (commonly called chalk) with some thin bands of marl and nodular chalk. There are also lines of nodular flints and sheets of tabular chert occurring throughout the White Chalk.

The flints commonly replace *Thalassinoides* burrows and so may have complex branching forms. But how were they formed? It is now thought that the silica was derived from sponge spicules and other siliceous planktonic organisms. Silica is secreted by these organisms and was soluble in the calcareous sediment at the bottom of the chalk sea. It would eventually sink to the sea floor and be precipitated within the numerous burrows excavated by arthropods and other neritic creatures. Marine life is represented by brachiopods, including rhynchonellids and terebratulids and echinoids such as the well-known 'heart urchin' *Micraster*, which is a useful zone fossil. Ammonites are common in the lower part of the chalk sequence and they are valuable for correlation purposes.

One of the most striking features of the chalk succession is the regular rhythmic repetition of marl-chalk units, nodular chalk beds, bands of flint nodules and layers of chert. Scientists have shown that the periodicity of these rhythms is around twenty to forty thousand years. These figures can be related to the Milankovitch cycles, which are produced by variations in the Earth's orbit and consequent variations in the amount of solar radiation. Thus, during the late Cretaceous, the cyclicity of sedimentation appears to be controlled by the Milankovitch climatic events though just how is uncertain.

Towards the end of the Cretaceous, the chalk was uplifted and raised above sea level to produce a land area that sloped gently to the southwest. Subsequent marine erosion planed the surface, removing some 25m of chalk at Culver in the east, while to the south of Alum Bay, in the west, the highest beds of chalk remained untouched. It was some 15 Ma before the youngest Palaeocene beds were laid down, hence we have a clearly defined unconformity marking the end of the Chalk Group and indeed extinction for many of the creatures that had dominated the Cretaceous world, including the dinosaurs and the ammonites.

Although the Chalk occupies a large area in the Isle of Wight, there are few accessible coastal sites where exposures can be viewed.

continued on p.40

The south face of the Needles headland near the western end of Scratchell Bay. The chalk has been highly compressed and recrystallized by monoclinal folding during the Alpine orogeny.

Itinerary 6: Compton Bay

Walking distance: 2km

Warning: this walk should be undertaken at low tide, and hard hats are recommended beneath the vertical cliffs.

There is a car park above Compton Chine [SZ 370851] from where you can follow the coast path southeastwards for a short distance before descending to the beach. The Ferruginous Sands outcrop in the cliffs to the west of the chine.

These weathered brown sandstones dip north at about 40 degrees and they are overlain by the Sandrock, consisting of white quartzose sands alternating with dark mudstones.

The red-brown sandstone of the Carstone completes the Lower Greensand sequence here. Gault Clay outcrops some 400m from the beach access point, marked by a cliff re-entrant, but it is much obscured by slumping. This dark-blue clay is superseded by transitional beds of sandy clays that lead into the Upper Greensand. This consists of buff-coloured silts and fine-grained sandstones with Chert Beds outcropping in the upper part of the formation. A small promontory marks the top of the Upper Greensand at SZ 366853, and this is a significant point beyond which you should only continue onwards on a falling tide, otherwise there is a substantial risk of being cut off by the flood tide.

The Lower Chalk begins with the Glauconitic Marl, just west of the promontory. It contains a fossiliferous sandstone with ammonites, bivalves and sponges. The overlying Chalk Marl consists of alternating bands of marl and limestone with a rich fauna of ammonites, including *Schloenbachia*, the bivalve *Inoceramus* and the echinoid *Holaster*. Next, the Zig Zag Chalk, with grey limestones and minor marl bands, leads up to the Plenus Marls at the top of the Lower Chalk. The basal member of the White Chalk is characterized by nodular beds with chalk pebbles that are exposed at beach level some 300m west of the promontory.

The base of the next sequence of beds is marked by a distinctive chalk horizon stained orange-brown by iron oxides and much bioturbated. This latter term refers to the extensive *Thalassinoides* burrows that indicate that this hardground surface marks a pause in sedimentation when burrowing organisms were particularly active. The northward dip of the chalk increases to 60 or 70 degrees farther along the coast towards Freshwater due to the influence of the Purbeck-Wight monocline.

At this point, return along the beach to Compton Chine. Do not continue into Freshwater Bay as there is no escape route if you are cut off by the rising tide!

RIGHT: **Cliffs formed of Lower Greensand near Compton Chine. Strata dipping northwest show Ferruginous Sands overlain by banded heterolithic sands and clays and Sandrock at the top of the cliff.**

BELOW: **Ferruginous Sands (Lower Greensand Group), dipping gently northwest, form the cliffs in Compton Bay. The white chalk cliffs that lie around Freshwater Bay are visible in the distance.**

Itinerary 7: Freshwater Bay

Walking distance: 1km

Freshwater Bay is located in what was formerly the upper reaches of the Western Yar, where the river had cut a gap through the chalk ridge between Tennyson Down in the west and Afton Down to the east. The river has now lost all of its headwaters due to relentless marine erosion, which even threatened to separate the western peninsular of the Isle of Wight in the early years of the twentieth century. This was due to the removal of shingle from the beach to make concrete for Fort Redoubt and other defensive works. A massive sea wall had to be constructed to prevent further marine encroachment.

There is a car park near to the centre of the bay at SZ 346857, which is a good place from which to start exploring the local chalk cliffs. At low tide on the western side of the bay, you can see the strata steeply dipping to the north. The fractured beds are in the lower part of the White Chalk and are brecciated due to stresses caused by the monoclinal folding. Above the sea wall, the eroded surface of the chalk is overlain by brown Pleistocene gravels deposited within the former Western Yar valley.

Walk over the flint shingle beach to the eastern side of the bay, where the beds of White Chalk can be correlated with those in the west along the east–west strike. In the southeast corner of the bay, low-angle shear planes caused by intense pressure cut across the steeply dipping bedding planes within the chalk cliffs. Where bands of flints are seen marking the steep dip, they curve near the top of the cliff, showing the monoclinal folding. However, the chalk is overlain by the Coombe Rock, a Pleistocene deposit of chalk rubble where frost-shattered chalk fragments slid downslope and accumulated as irregular layers of chalky debris.

The tops of the cliffs are covered with brickearth, a windblown periglacial deposit of silt, clay and sand sometimes referred to as loess. You will notice also that there are some solution pipes filled with brown brickearth that has sunk down from above.

LEFT: **Chalk cliffs along the eastern side of Freshwater Bay showing the steep northerly dip. The chalk is overlain by Coombe Rock and brickearth; the latter has filled solution pipes within the chalk.**

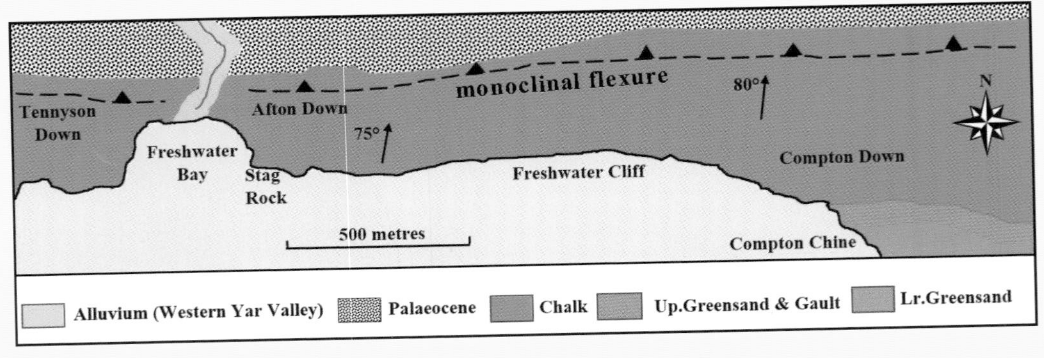

BELOW: **Geological map of the area around Compton Bay and Freshwater Bay.**

Mermaid Rock and Stag Rock, fine chalk stacks off the eastern headland of Freshwater Bay; Compton Bay lies beyond the stacks and the low cliffs of the 'Wealden coast' are seen in the distance.

Solution pipes are formed when the chalk is removed in solution by water moving through fissures and joints in the chalk. The sea stacks of Mermaid Rock and Stag Rock and the remains of Arch Rock stand off the eastern headland of Freshwater Bay. Arch Rock collapsed in October 1992, leaving two chalk stumps to mark its former position.

Finally, walk back towards the centre of the beach and stop near the wooden stairs to examine the Quaternary deposits that occupy the former valley of the Western Yar. Here the chalk, much fractured by freeze-thaw action, is overlain by fluvial sands that show evidence of cryoturbation. This is essentially the process by which the sediments are churned up by seasonal changes of temperature in a periglacial environment. Above the sands, there is a thick bed of gravel with flints and chalk debris (Coombe Rock), topped by a thin brickearth.

Pleistocene deposits in Freshwater Bay infilling the valley of the former Western Yar. Frost-shattered chalk is overlain by fluvial sands and Coombe Rock (flint and chalk debris), with a thin surface layer of brickearth. The sands have been churned up by freeze-thaw action under periglacial conditions.

Itinerary 8: Freshwater Bay to the Needles over Tennyson Down

Walking distance: 7km

This walk begins in Freshwater car park [SZ 346858], from where you can join the coastal path by taking the lane leading to Fort Redoubt and Watcombe Bay. From the clifftop above Watcombe Bay, you can look to the western face of the bay, where the Upper Chalk, with lines of flint along the bedding planes, dips steeply north. Coombe Rock overlies the chalk in the centre of the bay at the head of a small dry valley tributary to the Western Yar. This solifluction deposit was derived from frost-shattered chalk and flint that moved downslope under periglacial conditions in Devensian times. There is also a thin layer of brickearth covering the Coombe Rock.

The coast path rises gently over Tennyson Down to the viewpoint at Tennyson Monument and continues over West High Down before sloping down towards the New Battery, which was a rocket-testing site during the 1960s. Walk down to Sun Corner at the southwest end of the Needles peninsula, from where you can view Scratchell's Bay.

Here the Upper Chalk dips steeply northwards, with lines of flints marking the bedding planes (as at Watcombe Bay), and at the southern end of the narrow shingle beach, sea caves are being eroded at the base of the chalk promontory. St Anthony's Rock is an outlying sea stack, left isolated by the recession

ABOVE: **Scratchell's Bay and the chalk headland seen from a boat passing the Needles lighthouse.**

BELOW: **The Needles, chalk sea stacks eroded from strata dipping steeply to the north. The wider gap in the centre is where another stack, known as 'Lot's wife', disappeared in a winter storm in 1764.**

The Needles headland provides an excellent viewpoint, overlooking Alum Bay and beyond to Headon Hill.

of the cliffs. Walk up to the Coastguard Cottages for a view of the Needles and three stacks with the lighthouse at the seaward end. Another stack, known as Lot's Wife, collapsed in a storm in 1764 but its stump can still be seen at low tide.

You may well ask why the remaining stacks are still standing, given the relentless pounding they receive during winter storms. The answer is that the chalk is extremely hard in this location due to compression and recrystallization during the Alpine earth movements, which produced a northern dip of 75 to 80 degrees in the centre of the monoclinal fold.

While viewing the Needles headland and stacks from the south side, it is well to remember that bedding planes appear to be sub-horizontal or gently dipping; this is in fact almost on the strike direction, while the true (much steeper) dip shows on the western faces.

When you are on the Needles headland, it is worth visiting the Old Battery, which was built in the 1860s to defend against French invasion. It was never used at the time and became one of 'Palmerston's follies', although it was recommissioned as a signal and observation post during both world wars. The property is now under the control of the National Trust. On the north side of the Needles headland, overlooking Alum Bay, the highest beds of the Upper Chalk are less resistant to weathering and so the highly inclined beds produce steep cliffs. From here there is a marvellous view across the Palaeogene sands and clays of Alum Bay cliffs towards the headland, dominated by Headon Hill.

The coast path can be followed along to the car park at Alum Bay [SZ 306854] if return transport has been arranged.

Selected Fossils from the Island

Fossil hunting can be an absorbing hobby, but we should all be aware that many coastal sites on the Isle of Wight are under constant attack from the waves and landslipping and so we must respect the geological environment and refrain from hammering indiscriminately. Many sites have SSSI status, which means that the bedrock cannot be disturbed in any way and only specimens found in loose rocks on the foreshore may be collected. It is far better to photograph fossils than remove them from their natural habitat! You can build up an indexed collection of photographs that will take up less room than bulky trays of 'real' fossils.

Wealden Fossils

Viviparus is a common freshwater gastropod found at various horizons in the Wealden Group. The Wessex Formation is best known for its dinosaur footprints and bones, particularly the remains of *Iguanodon*. Water-worn reptilian vertebrae can sometimes be found on the foreshore along Compton Bay. The Vectis Formation has several common examples of invertebrate fossils, including the freshwater mussel *Pseudunio valdensis* and the brackish water bivalve *Filosina*. The latter is abundant in three thin limestone bands near the top of the Vectis Formation. The oyster *Ostrea* occurs in an overlying limestone and marks the approach of marine conditions at the end of Wealden times.

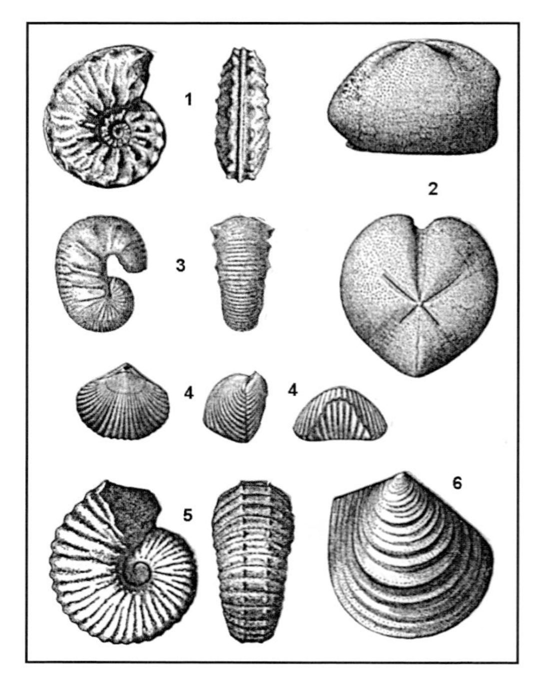

Fossils from the Chalk: 1. *Schloenbachia* (ammonite); 2. *Micraster* (echinoid); 3. *Scaphites* (ammonite); 4. *Orbirhynchia* syn. *Rhynchonella* (brachiopod); 5. *Mantelliceras* (ammonite); 6. Inoceramus (bivalve). Not to scale.

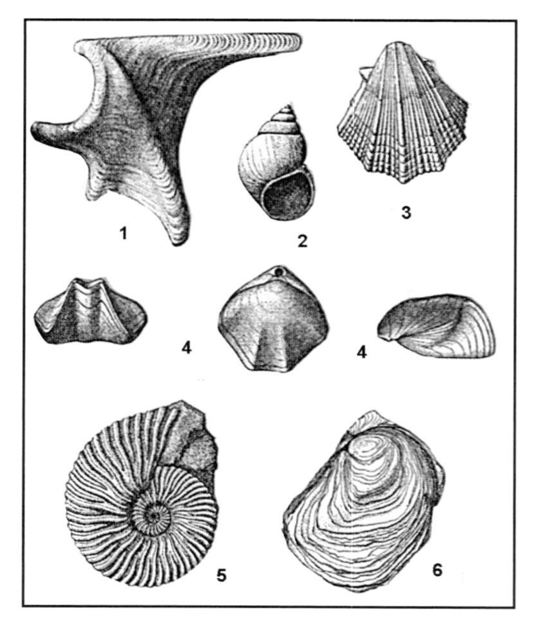

Fossils from the Wealden Group and Lower Greensand: 1. *Mulletia* syn. *Perna* (bivalve); 2. *Viviparus* (gastropod); 3. *Neithea* syn. *Pecten* (bivalve); 4. *Sellithyris* syn. *Terebratula* (brachiopod); 5. *Deshayesites* (ammonite); 6. *Aetostreon* syn. *Exogyra* (oyster). Not to scale.

Lower Greensand Fossils

Mulletia (formerly *Perna*) is a large bivalve that gives its former name to the Perna Bed at the base of the Atherfield Clay. This formation also contains the small lobster *Meyeria magna* and the strongly ribbed ammonite *Deshayesites* from the concretions in the Crackers Bed. The

Ferruginous Sands are rich in fossils, including the large oyster *Aetostreon* (formerly *Exogyra*) and the distinctive partially uncoiled ammonite *Australiceras*. Other ammonites are *Tropaeum*, *Cheloniceras* and *Parahoplites*. Large concretions in the Ladder Chine Beds contain colonies of the terebratulid *Sellithyris* (brachiopod).

Chalk Fossils

The Chalk Marl is characterized by ammonites such as *Schloenbachia*, with its strong ribs and prominent keel, and the involute ribbed *Mantelliceras*. The strange ammonite *Turrilites* has its shell coiled in a helicoid spiral. The bivalve *Inoceramus* and the echinoid *Holaster* are also found in the Lower Chalk. *Micraster coranguinum*, better known as the 'heart urchin' and *Echinocorys* are common echinoids in the White Chalk. Brachiopod shells include the rhynchonellid *Orbirhynchia* and the terebratulids *Gibbithyris* and *Terebratulina* that are usually concentrated in certain horizons within the White Chalk.

Fossils from the Palaeocene: 1. *Planorbina* (gastropod); 2. *Nummulites* (foraminifera); 3. *Galba* (gastropod); 4. *Turritella* (gastropod); 5. *Megalocochlea* (gastropod); 6. *Venericor* (bivalve); 7. *Athleta* (gastropod). Not to scale.

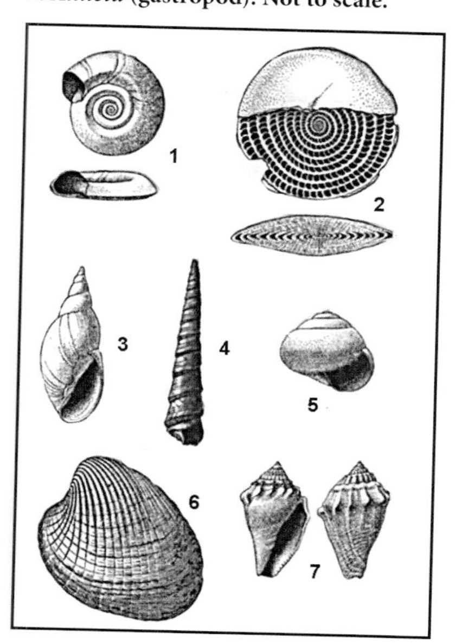

Palaeogene Fossils

The London Clay is one of the most fossiliferous Palaeogene beds, containing the gastropod *Turritella* and bivalves *Pholadomya*, *Panopea* and *Glycymeris*. *Venericor* is a distinctive ribbed bivalve from the Bracklesham Beds, and the overlying Barton Clay has a rich molluscan fauna including *Crassatella*, *Corbula*, *Chama* and *Athleta*. The Headon Hill Formation also has many freshwater water gastropods, including *Potamides*, *Batillaria* and *Galba*. The last of these also occurs in the Bembridge Limestone, along with other freshwater gastropods such as *Palaeoglandina* and *Planorbina*. Finally, the Hamstead Beds yield a range of gastropods, including *Athleta* and *Potamaclis*.

What are Sedimentary Rocks?

When existing rocks are weathered and eroded, the resulting material is transported and deposited as sediment in rivers or lakes, on the sea floor or on the surface of the land. Lithification is the process by which unconsolidated sediment is converted into a sedimentary rock. The sediment is first compacted by the weight of overlying sediment and water is squeezed out of the pores between the grains. Silica or calcite may be precipitated from solution to form a cement around the grains. Iron oxides also form cement and give the characteristic red colour to sandstones. Porosity is the amount of pore space between grains or in fissures. By contrast, permeability is a measure of fluidity – that is, the ability of fluid to move through interconnected pore spaces.

Sedimentary rocks: fine sandstone above ripple drift siltstone.

There are two main groups of sedimentary rocks: clastic rocks, which are formed mostly of quartz grains or lithic fragments (clasts); and carbonate rocks, which are formed by the precipitation of carbonate minerals in shallow seas.

Clastic Rocks

These are classified according to grain size.

Mudstones are exceedingly fine-grained, being composed of clays with a particle size of <0.004mm and silts 0.004–0.062mm.

Shale is the term used for fissile mudstones.

Sandstones range from 0.062mm to 2mm; gravels 2–4mm; pebbles 4–64mm; cobbles 64–256mm and boulders >256mm.

Conglomerates contain rounded clasts while breccias contain angular clasts.

Turbidites are sediments deposited by turbidity currents off the continental shelves and may contain a range of grain sizes.

Carbonate Rocks

These form in shallow shelf seas as carbonate sediments. They are accumulating today in the Arabian Gulf and off the Bahamas, but they were more common in past geological periods.

Bioclastic limestones are calcareous rocks formed largely of the remains of marine animals such as bivalves, brachiopods and corals. The clasts are often cemented together by calcite. These limestones were deposited as sediments on offshore ramps in shallow waters, particularly in Silurian and Carboniferous times.

Oolitic limestones are formed of tiny, concentrically layered carbonate grains called ooliths. The calcium carbonate is precipitated in shallow shelf seas under tropical conditions. Today ooliths are being deposited in the Persian Gulf and off the coast of Florida, but in the past oolitic limestones were formed mainly during the Jurassic period.

Magnesian limestone (Dolomitic limestone) is formed of the mineral dolomite or calcium magnesium carbonate. Most dolomitic limestones are formed by the replacement of the calcium ions by magnesium ions in the calcareous sediment when water evaporates. During Permian times, dolomite was deposited as an evaporate mineral in the Zechstein Sea.

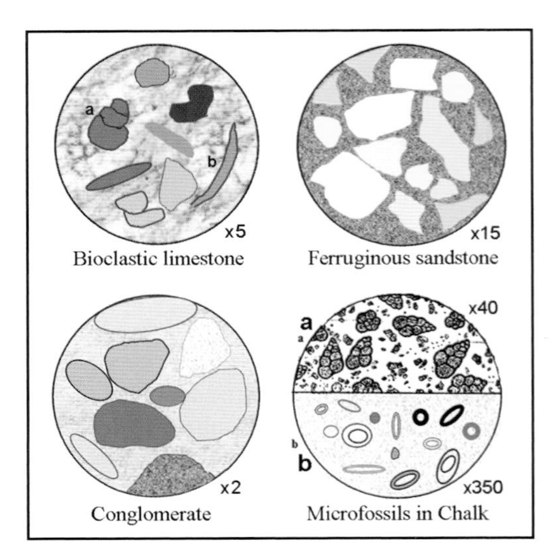

Textures of some typical sedimentary rocks, as seen when thin sections are viewed under a petrological microscope.
The bioclastic limestone has a calcite matrix, with sections of (a) gastropod and (b) bivalve fossils, as in the Bembridge Limestone.
The ferruginous sandstone has sub-angular quartz grains with hematite cement, as is common within the Lower Greensand.
The conglomerate has flint pebbles in a sandy matrix, as in the Reading Beds.
The microfossils in Chalk are foraminifera (at x40) and coccoliths (at x350).

Chalk is a very pure form of limestone made from microscopic skeletal plates (coccoliths), which accumulated as a carbonate-rich mud on the seabed in Upper Cretaceous times. The Upper Chalk is characterized by the presence of flint nodules that may have been precipitated from silica-rich groundwaters percolating through the chalk. However, recent research suggests that the flint was formed by the sub-surface breakdown of siliceous organisms such as sponges, radiolaria and diatoms during the deposition of the chalk.

CHAPTER 5

Palaeogene: the Sea Comes and Goes

The end of the Cretaceous was marked by one of the greatest mass extinctions of all time, when three-quarters of all living creatures, including the dinosaurs and the ammonites, vanished from the face of the Earth. The asteroid impact in the Gulf of Mexico 66 million years ago is popularly blamed for this mass extinction event but it was probably caused by a combination of factors, including plate movements, global temperature and sea level changes and volcanic eruptions, in addition to asteroid collision.

After these catastrophic events, the land surface again became vegetated with trees, ferns, flowers and grasses, providing food for the evolving and diversifying mammals. In the seas, the bivalves flourished alongside fish and crustaceans, and the climate on the Isle of Wight was warm and temperate.

The Palaeogene extended from 66 Ma to 23 Ma; and there are three divisions – the Palaeocene, Eocene and Oligocene. In older literature you will find that the Palaeogene is referred to as the Lower Tertiary. During this time, the Hampshire Basin (including the Isle of Wight) was a region of undulating low relief that was drained by rivers flowing off the higher land to the northwest. These rivers brought down gravels, sands and silts that were deposited on floodplains and deltas and in estuaries and lagoons. However, from time to time the sea level rose and marine transgression occurred, covering the land with finer sediments.

The Palaeogene strata show evidence of several cycles of sedimentation; each cycle consists of marine sediments deposited as the sea advanced from the east, followed by terrestrial sediments deposited by rivers advancing from the west. Significantly, the Isle of Wight lies roughly in the centre of the Hampshire Basin, and so in Whitecliff Bay, on the east coast, there are largely marine sediments, while in Alum Bay, on the west coast, there are exposures of mainly terrestrial sediments.

The northern half of the Isle of Wight is formed of Palaeogene strata that generally display sub-horizontal bedding, except on the northern side of the east–west monoclinal fold, where steep, almost vertical dips occur.

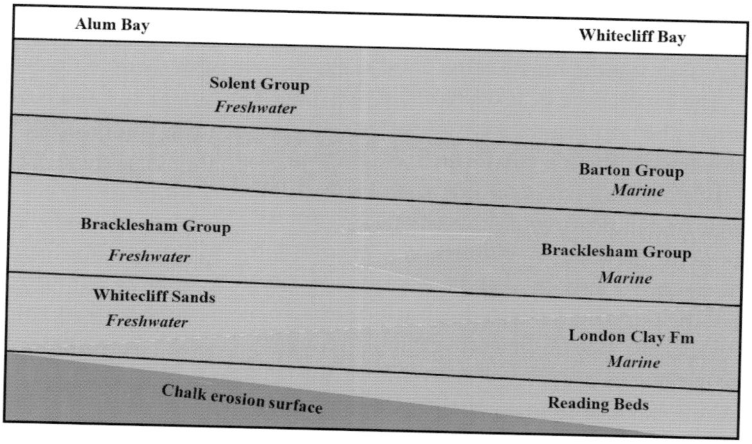

Diagram showing the cycles of sedimentation within the Palaeogene sediments that crop out across the northern half of the Isle of Wight.

Palaeogene Strata by Age

Let us first consider the Palaeocene sediments known as the Reading Beds, which were originally described and named in the Thames valley. A thin basal conglomerate and a sand bed were deposited on the eroded surface of the Chalk above the Cretaceous unconformity as the sea advanced from the east across the Hampshire Basin. Yet the marine episode did not last long and, as the sea regressed, freshwater muds were laid down in rivers and shallow lakes. Due to their steep dip, the Reading Beds form a narrow outcrop along the northern margin of the central chalk ridge.

This first cyclic sequence is incomplete in the Isle of Wight since the truly marine Thanet Sands of the London Basin are missing. However, as we move upwards into the London Clay Formation, there is evidence of considerable marine influence. Gastropods such as the well-known *Turritella* and the bivalve *Panopea* are characteristic marine fossils from the London Clay. Nevertheless, towards the end of London Clay times, there was an influx of coarse deltaic sediment brought down by rivers from the western land area. These are the Whitecliff Sands (formerly known as the Bagshot Beds), which consist of cross-stratified quartz sands of varying colours – orange, brown or red depending on the amount of iron staining. There are also some seams of pipe clay, which formed in shallow pools on the delta top. The pipe clay is so called because it was used in making clay tobacco pipes during the nineteenth century.

The succeeding Bracklesham Group (named after the exposures near Bracklesham on the Sussex coast) are formed of glauconitic sands and muds containing marine molluscs that are indicative of warm, shallow waters. Foraminifera such as *Nummulites* are common at certain horizons and have been used for zoning purposes. Whilst the Bracklesham Beds on the east side of the Isle of Wight are mostly marine, it is a different story in the west. In Alum Bay the sands and silts are interpreted as fluvial and lagoonal in origin and there are repeated layers of lignite (plant debris) and palaeosols (fossil soils).

The overlying Barton Group is composed almost entirely of marine sands and clays. The type area for these beds is Barton-on-Sea on the Hampshire coast. There is a diverse prolific fauna of marine bivalves and gastropods that indicate a gradual increase in the depth of water until near the end of Barton times, when shallow beach conditions returned, depositing clear quartz sands. Some horizons are crowded with fossils, including the bivalves *Chama*, *Crassatella*, and *Corbula* and the gastropods *Turritella* and *Athleta*.

The Solent Group occurs at the top of the Palaeogene strata in the Isle of Wight and was laid down almost entirely under terrestrial conditions except for a brief interval at the beginning of Solent times. The Headon Hill Formation is named from the hill above Alum Bay, and its sands and clays, which are exposed in the cliffs of Totland Bay and Colwell Bay, contain molluscs that lived mostly in an estuarine and lagoonal environment. The overlying Bembridge Limestone appears to have been deposited in freshwater lakes that formed as the sea regressed eastwards. A variety of freshwater gastropods such as *Megalocochlea* and *Palaeoglandina* are commonly found in these calcareous beds.

The outcrop of the Bembridge Limestone can be traced across the island from above Sconce Point (Fort Victoria Country Park) in the west to Bembridge Foreland in the east. The youngest sequence of beds on the Isle of Wight is known as the Bouldnor Formation, which comprises the Bembridge Marls and the Hamstead Beds. The former represent a brackish water lagoonal environment in which silt and clay accumulated. Seeds and plant remains have been found, which suggest a subtropical type of vegetation. There is also a fine-grained calcareous mudstone that has yielded a variety of well-preserved insects.

continued on p.53

Itinerary 9: Alum Bay

Walking distance: 2km

The car parks nearest to Alum Bay are beside the Needles Pleasure Park [SZ 306854] but during the height of the tourist season, both the car parks and the beach can be very crowded. You can access the beach via the steps down Alum Bay Chine or take the chair lift when it is operational in the summer.

Before descending to the beach from the cliff-top it is worth viewing the Needles, which extend to the western tip of the island. The line of sea stacks is formed of a hard compact layer within the White Chalk Formation that dips at 80 degrees north. When you get down to the beach it is best to walk south to the base of the chalk cliff and then work back through the steeply dipping Palaeogene section. This means that you will be progressing stratigraphically upwards!

ABOVE: **Alum Bay Chine: like most of the chines on the Isle of White chines, this was cut a by short, fast-flowing stream with a steeply graded long profile.**

ABOVE: **North side of Alum Bay, where almost vertical Bracklesham Beds are overlain by Barton Clay in Alum Chine, where the chairlift is located.**

BELOW: **Cliff profile across Alum Bay and Headon Hill.**

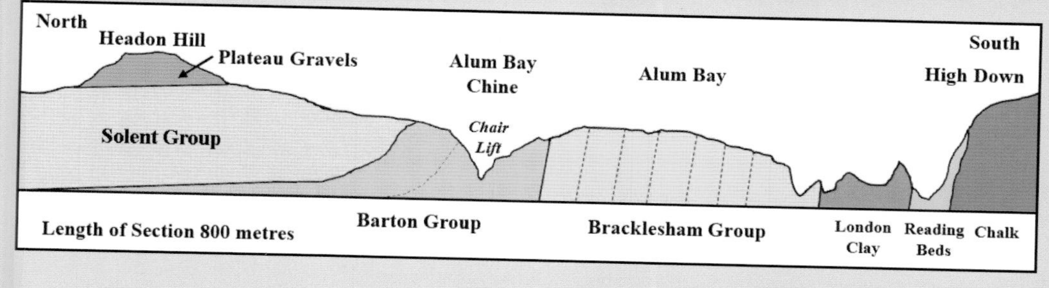

LEFT: **The eroded clifftop in Alum Bay on the outcrop of the Bracklesham Beds. The steeply dipping iron-rich band stands proud of the clays on either side.**

BELOW: **Deeply eroded, soft and unstable clays at Alum Bay produce a 'badlands' type of topography. Clifftop recession takes place as a result of downwash during heavy rain. The leaning wooden post in the right of the photograph is an indication of creeping surface movement.**

The boundary between the Chalk and the Reading beds is normally visible in the southern corner of the bay, marked by a deep valley alongside the wall of chalk. This is an unconformable junction and you will notice that there are some solution hollows up to 2m in diameter within the almost vertical chalk surface. These potholes would have been cut into a horizontal subaerial surface at the end of the Cretaceous before the basal Reading Beds were deposited. Subsequently, in Miocene times, powerful tectonic forces upended the strata creating the steep dips that we see today.

BELOW: **Cliffs of Alum Bay expose Barton Beds on either side of the chine and chairlift. Bracklesham Beds extend to the right in the photograph, and are nearly vertical on the steep limb of the monocline. The dip becomes less towards the north, where the Headon Hill Beds on the left are almost horizontal.**

Here the Reading Beds are coarse ferruginous sands containing flint nodules derived from the chalk. Unfortunately, a mudflow obscures much of the sequence in this locality. The succeeding London Clay begins with a glauconitic sandy clay known as the Basement Bed, followed by three coarsening upwards cycles of sedimentation, reflecting marine transgression followed by shallowing. The so-called Whitecliff Sands form a narrow outcrop at the top of the London Clay succession.

The Bracklesham Group is about 230m thick and occupies the centre of the Alum Bay section. It consists of sand beds that alternate with heterolithic units of muds, silts and sands. The cliffs are higher and less dissected here, and the harder, more cemented sand bodies tend to form protruding ribs separated by gullies in the softer clay bands. It is the sand beds that produce the famous Alum Bay coloured sands. White sands are pure quartz grains with no mineral staining. Green sands are formed by the presence of glauconite but, when weathered, this mineral breaks down into various iron oxides. Haematite produces red colours, limonite forms brown and yellow colours and black sands are formed by the presence of finely comminuted lignite. Several plant beds occur in the middle of the Bracklesham Beds, including the renowned Alum Bay Leaf Bed, which is a white pipe clay containing leaf impressions.

Unconformable junction between the Reading Beds and the White Chalk in Alum Bay. The erosion surface (now steeply inclined) contains large solution hollows that are filled with brown sands of the Reading Beds, which are about 15 million years younger than the chalk. Some flints from the chalk have been incorporated in the infill.

The overlying Barton Group begins about 30m south of the chair lift where the Alum Bay Chine is cut into the Barton clays. North of the chine you will notice an abrupt change of dip where the strata become almost horizontal as Headon Hill is approached. This change may be due to faulting, although it is difficult to see the fault plane.

The southeast corner of Alum Bay, with Bracklesham Beds in the left foreground, and slumped London Clay, (dark brown) and Reading Beds resting unconformably on the steeply inclined White Chalk of West High Down.

Itinerary 10: Whitecliff Bay and Culver Down

Walking distance 4.5km

Access is via Hillway Road near Bembridge airport. Turn into the Whitecliff Bay Holiday Park, where it may be possible to park cars [SZ642864]; otherwise there is a public car park on Culver Down [SZ636856]. A zig-zag track leads down to the beach in the centre of the Whitecliff Bay.

Walk to the southwestern end of the bay, where the chalk bastion of Culver Down stands proud. Here the unconformable junction with the Reading Beds is rather obscured by slumping at beach level. The steeply inclined chalk surface is potholed and worn, reflecting the fact that some 25m of chalk were removed by erosion before the deposition of the Reading Beds. A thin flinty conglomerate forms the base of the Reading Beds, which consist mainly of unfossiliferous red muds that are much eroded by downwash.

The slumped cliffs of the overlying London Clay Formation contain several coarsening upward cycles resulting from fluctuating sea levels. In the upper part of the sequence look for lines of septarian nodules. These were probably formed shortly after deposition, when calcareous iron-rich clay segregated into compact rounded masses. Later, a complex array of concentric and radial shrinkage cracks developed in which calcite and siderite were precipitated. When cut open and polished, septarian nodules can display beautiful crystalline structure. The London Clay also has a range of common marine bivalves, including *Pholadomya*, *Pinna* and *Glycymeris*.

As the sea regressed at the end of London Clay times and rivers brought down more sediment, cross-bedded sands were deposited in estuaries and tidal channels to form the Whitecliff Sands, previously called the Bagshot Sands.

LEFT: **The south-western end of Whitecliff Bay, showing the slumped cliffs of London Clay, the yellow Whitecliff Sands and part of the Bracklesham Beds.**

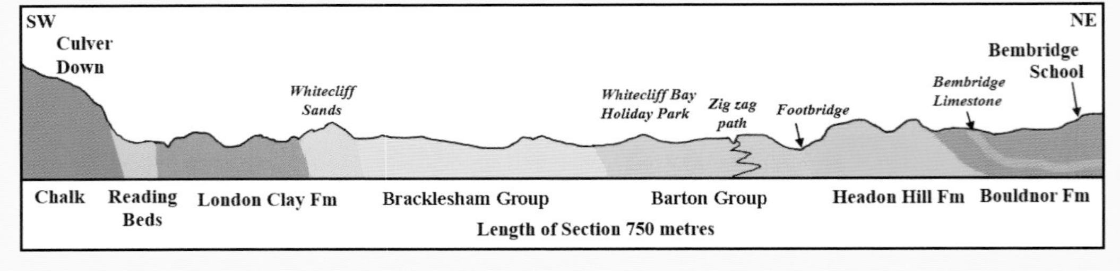

BELOW: **Cliff profile across Whitecliff Bay from Culver Down to Bembridge School.**

ABOVE: **The view from the crest of Culver Down across Whitecliff Bay and beyond to the Foreland.**

BELOW: **Cliff exposure of Whitecliff Sands (formerly known as Bagshot Sands) in Whitecliff Bay. This buttress of vertically bedded sandstone is in marked contrast to the gulley on the left that is cut into much softer London Clay.**

Rich Fossil Beds

The Bracklesham Group forms a cyclic sequence of sands and clays in the centre of the bay. Near to the base of this group is the Cardita Bed, a type of fossilized cockle bed that contains abundant specimens of *Venericor*. The 'screw shell' *Turritella* is also commonly found in upper part of the Bracklesham Group. About 100m from the zig-zag path, near the sea wall, you should look for a shelly bed containing numerous specimens of *Nummulites laevigatus*, a large species of foraminifera 1.5cm in diameter. These are large, flat shells like coins, hence the name (Latin *nummus*, a coin); if you split them with a penknife, you will see a spiral of tiny chambers. During the Eocene, vast beds of nummulitic sediments were deposited in the Tethys sea, creating a distinctive limestone from which the pyramids of Egypt are constructed.

The overlying Barton Group outcrops on either side of the zig-zag path, although the lower beds of Barton Clay are not well exposed due to slumping. Immediately to the north of the path is the Chama Bed, so named for its characteristic bivalve *Chama*. This is followed by the buff-coloured Becton Sand, which forms a vertical cliff.

The junction between the Becton Sand and the Headon Hill Formation is seen near to the path leading down the gully from the footbridge. Here there is a rapid change from the marine sands to muddy sediments containing pond snails *Viviparus*, *Galba* and *Planorbis* that lived in shallow freshwater lakes; other molluscs were adapted to life in brackish water lagoons.

The succeeding Bembridge Limestone is well exposed on the north side of the bay. It is formed of several limestone beds with intercalated clays that dip steeply northwards before levelling out and following the shoreline northwards. At Black Rock Point the limestone extends seawards across the foreshore to form a series of ledges, and these continue around the Foreland as the Bembridge Ledges. The Bembridge Limestone is packed with the moulds of freshwater snails (*Galba* and *Planorbis*) and with an aquatic plant known as *Chama* that is set in calcareous matrix. The Bembridge Marls of the Bouldnor Formation rest on the limestone beds and form much of the cliffs below Bembridge School. They consist mostly of marls and thin sands but at the base there is an oyster bed that is exposed at Black Rock Point.

Up on the Downs

Return to the car park and, if time permits, why not make for the Culver Haven pub [SZ 633855] for light refreshment before tackling the heights of Culver Down. Near the summit stands a monument to the memory of the Earl of Yarborough, erected in 1849 and built of vermiculated granite blocks imported from southwest England at great expense! There are fine views in all directions from here: to the north across Bembridge airport towards Bembridge harbour; to the northeast across Whitecliff Bay and the Foreland; and to the southwest over the Lower Greensand lowlands reaching the coast at Red Cliff and Sandown. Westwards the chalk hog's back extends across Bembridge Down, the Yar gap and to Brading Down and beyond.

From the monument, follow the coast path down the southern slopes of the chalk ridge to where it meets the cliff edge and then continue along the cliff top until it descends into a steep re-entrant that cuts across the head of the Yaverland dry valley at SZ 626855. At this point you will have descended off the Chalk and Upper Greensand, and the break of slope marks the junction with the narrow outcrop of the Gault Clay that the dry valley follows towards Yaverland. Now look eastwards along the coast towards Culver Cliff and notice the slumped Gault Clay and Upper Greensand with the chalk cliffs beyond. These cliffs rise above the wave-cut platform known as Whitecliff Ledges. Since the strata dip steeply north, it is only the Lower Chalk that is exposed on the southern side of the headland at Culver Cliff; the higher beds outcrop around Whitecliff Point.

Continue along the coast path that rises over Red Cliff, where the Ferruginous Sands reach the coast. Beyond the cliff is the landslipped zone of the Atherfield Clay. At Red Cliff you should leave the coast path and follow the track that crosses the dry valley and climbs up the chalk slope of Bembridge Down. The summit of the ridge is crowned by Bembridge Fort; this was built in the 1860s as one of the Palmerston forts to counter the threat of French invasion, but they soon became obsolete. Today the building is owned by the National Trust and is open to the public. The Yarborough monument was originally sited here but was moved to Culver Down stone by stone to make way for the fort. Return to the car park along the road running along the crest of the ridge.

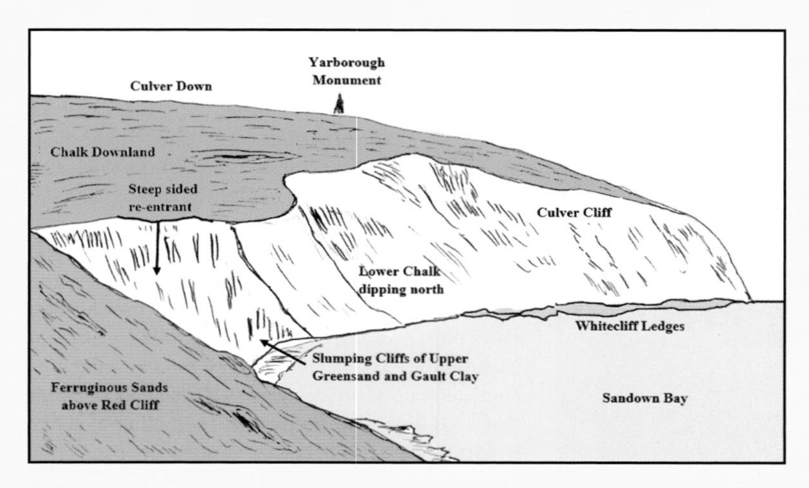

Field sketch looking east across the northern end of Sandown Bay to Culver Cliff and the summit of Culver Down, both of which are formed of northwards dipping Chalk.

Itinerary 11: Hamstead Cliff

Walking distance: 3km

Although the Bouldnor Formation covers a large area in the northern half of the Isle of Wight, there are few coastal sections where the whole sequence can be examined. The best locality is on the foreshore and cliffs southwest of Hamstead Point, where both the Bembridge Marls and the Hamstead Beds are exposed.

Access from the A3054 Yarmouth–Newport road is via a track leading to Hamstead Farm. There is parking space for a few cars just before the farm is reached at SZ 401911. Join the coast path and walk towards Hamstead Point [SZ 406921], descending to the foreshore at Hamstead Ledges.

These beds of Bembridge Limestone, separated by softer clays, extend seawards from below the Bembridge Marls at the base of the cliff. The ledges dip gently southwards since they are on the northern flank of the Bouldnor syncline. In the cliffs just above the top of the limestone there is a thin shell bed containing *Ostrea* but the lower Bembridge Marls mostly contain brackish water molluscs such as *Corbicula* and *Melanoides*. The upper part of the marl sequence is quite fossiliferous and the frequent occurrence of *Viviparus* suggests a freshwater environment. About 100m along the foreshore southwest of the ledges there are several carbonaceous seams that contain the fruits, seeds and leaves of aquatic plants that were probably growing in freshwater lacustrine conditions.

Now look out for a line of posts running out to sea. About 200m before you reach the posts, there is a distinctive black carbonaceous bed about 0.75m thick that marks the base of the Hamstead Beds. These beds are significant in that they are Lower Oligocene in age and they are found nowhere else in the Hampshire Basin. When you reach the line of posts, work your way slowly past slumped clays where intermittent exposures of the Hamstead Beds will reveal shelly bands containing a range of freshwater molluscs such as *Galba*, *Unio*, *Viviparus* and *Planorbis*.

Continue for some 750m beyond the posts to where a bed of fragmented white shells interspersed with mud bands is visible. There are also some bands of tabular clay ironstone, which was collected and sent to Swansea for smelting in the nineteenth century. A few metres further on is the Waterlily Bed, a carbonaceous mud with plant debris; the leaves and seeds of water lilies and palm trees, together with the scattered remains of turtle and crocodile, suggest a brackish water lagoonal environment with a subtropical climate. After exploring this fascinating shoreline, you can return to Hamstead Farm by taking the path up the cliffs from the centre of the bay.

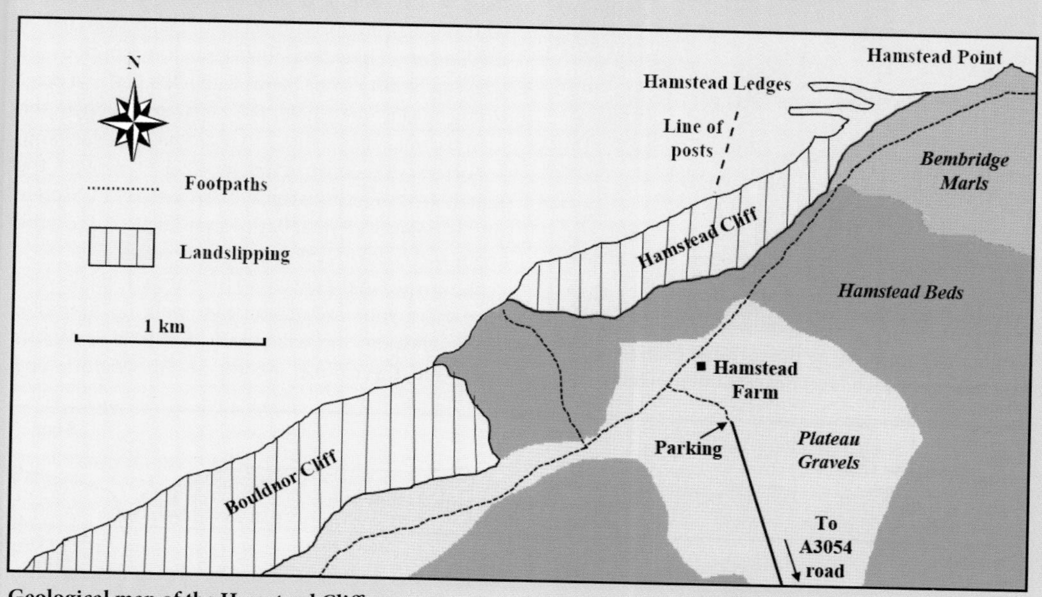

Geological map of the Hamstead Cliff area.

Itinerary 12: Colwell Bay and Sconce Point

Walking distance: 2km

This is a short walk around Colwell Bay, starting at the car park at Colwell Chine Road [SZ 327878].

The strata exposed along the cliffs and on the shore belong to the Headon Hill Formation and the lowest division forms the Totland Bay Beds, which are represented by a calcareous sandstone that extends seawards as a rock platform at Warden Ledge, dipping gently northwards.

Immediately north of Colwell Chine is the succeeding How Ledge limestone, exposed in the low cliffs behind the sea wall, and this bed dips down to beach level at How Ledge about 200m beyond the chine. The light-brown limestone contains freshwater gastropods such as *Galba* and *Planorbina*.

As you approach Brambles Chine, the Colwell Bay Beds, consisting of shelly sandy mudstones, outcrop at the foot of the cliffs. One particular unit, known as the Venus Bed, is very fossiliferous, with a varied assemblage of bivalves and gastropods. The bed is named after the characteristic shell *Venus*, a common marine bivalve that burrows into shallow water sediments. The bivalves are often found in life position; but in this locality there is also an overlying oyster bed containing groups of *Ostrea velata*.

Bivalves in a sandy matrix within the Colwell Bay Beds near Brambles Chine.

Sandy marls containing the turreted gastropod *Potomides* at beach level near Brambles Chine.

The upper part of Brambles Chine is cut into the Linstone Chine Beds, which dip northwards to beach level until just beyond Linstone Chine, where a small anticline raises them to the cliff top. However, along most of the section, the overlying Cliff End Beds rest with a sharp erosional contact on the sands and clays of the Linstone Chine Beds. Bivalves and gastropods are again common in shell bands within the Cliff End Beds. North of Linstone Chine, the cliffs are slumped and irregular, making it difficult to distinguish the bedding.

Fort Albert, constructed in 1856 to protect the Solent from French attack, soon became obsolete and is now residential flats. Hurst Castle spit on the mainland is only 1.25km across the water.

Return along the beach to the car park. If you wish to extend your excursion, then walk up Brambles Chine and join the coast path that follows Monk's Lane and the clifftop leading to Fort Victoria and

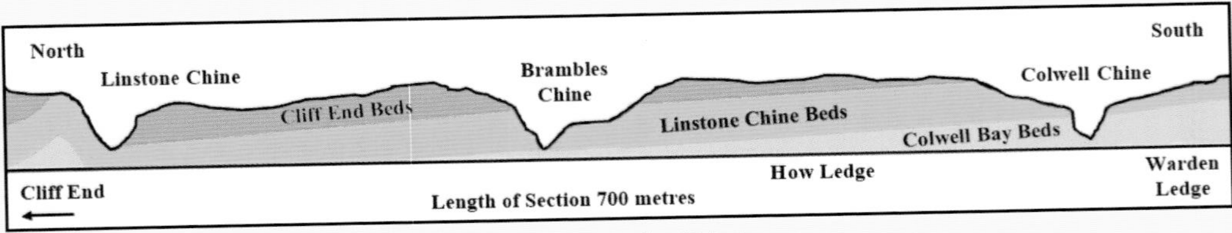

Section across Colwell Bay from Linstone Chine to Warden Point.

Sconce Point. This section of the coast is formed of the Osborne marls at the top of the Headon Hill Formation and they form slumped unstable cliffs that are only easily accessible at Sconce Point, where shelly mudstones containing *Viviparus*, the freshwater snail, are seen at beach level. The shore is also littered with boulders of Bembridge Limestone that have fallen from its outcrop near the top of the cliffs. Note that the limestone does not form reefs or ledges here as in other parts of the island, since it overlies the Osborne beds well above sea level.

The coast path follows the outcrop of the Bembridge Limestone on the wooded ridge that runs parallel to the coastline from Cliff End to Norton village beyond Sconce Point. Fort Victoria Country Park has been established in recent years on land overlooking Sconce Point.

ABOVE: **Anticlinal fold where iron-rich sands and clays of the Linstone Chine Beds are exposed. Plant remains found in these beds suggest a vegetated estuarine environment.**

LEFT: **The Cliff End headland with Fort Albert at the northern end of Colwell Bay.**

continued from p.44

Finally, the Hamstead Beds that are exposed on the crumbling Bouldnor cliffs are again largely fresh water in origin with a varied molluscan fauna. These clays and sands are the youngest bedrock to be exposed on the island. It is possible that further deposition occurred, but if so the material evidence has been removed by erosion.

If you have access to the British Geological Survey 1:50,000 Isle of Wight Special Sheet, you will notice that the Solent Group of rocks, and particularly the Hamstead Beds, cover much of the northern half of the Isle of Wight. This is not related to the thickness of the beds but rather to the low, mostly horizontal dip of the strata. Also, if you look at the map key to the Palaeogene rocks, you will see that the Eocene/Oligocene boundary is placed at the junction between the Bembridge Limestone and the Bembridge Marls. In the past, many authors considered the boundary to be drawn at the base of the Headon Hill Formation but more recent palaeontological evidence has resulted in the boundary being revised upwards, although it is still not agreed as an international standard. You should note that although the Oligocene is relatively thin in southern Britain, the Bouldnor Formation is about 100m thick, and it is much more highly developed in continental Europe.

Building Stones

Many of the older buildings on the Isle of Wight reflect the local geology. Although there are predominantly soft sands and clays on the island, there are also more resistant beds at various horizons. Within the Lower Greensand, the hard bands in the Ferruginous Sands provide useful building material, and the sandstone beds of the Carstone have been used in the construction of Mottistone Manor and other older buildings and churches around the Brighstone area.

Along the south coast, from Bonchurch to St Catherine's Point, masses of Chalk and Upper Greensand have slumped down over the Gault to form the Undercliff. Yet the most significant of these rocks, from a building perspective, is the Upper Greensand, which emerges from beneath the chalk above the Undercliff. It also forms a wide marginal outcrop around the chalk downs. Near the top of the greensand sequence is a thick freestone bed of fine-grained sandstone known as the St Boniface Stone, which forms an excellent building stone that has been used in local churches, barns and farm cottages. The succeeding Chert Beds are hard siliceous sandstones with multiple layers of chert that are used for local building, but are best seen in Gore Cliff near Rocken End.

Bembridge Limestone is a hard, fine-grained freestone much used by medieval masons for ecclesiastical buildings. Here it forms the cliffs on the north side of Whitecliff Bay.

On the chalk hills, the availability of flint nodules and hard bands of chalk, such as the Melbourne Rock, provided building material for village churches, farmsteads and field boundary walls. Numerous shallow pits can be seen on the downlands that supplied the chalk for both building and lime production.

The Bembridge Limestone is the best known building stone on the Isle of Wight. It has been quarried around Binstead near Ryde since Roman times, when it was used in the construction of Portchester Castle. In the twelfth century, the Cistercian monastery at Quarr was built of the so-called Quarr Stone, but after the dissolution of the monastery the stone was reused to build Yarmouth Castle when Henry VIII decided to fortify the Solent approaches. Quarr Stone was mostly exported to the mainland but, being a lenticular deposit of limited extent, it was largely worked out by the end of the fifteenth century. Bembridge Limestone was also quarried extensively in the coastal outcrops near St Helens and was used in building Winchester Cathedral, Christchurch Abbey and in the medieval arches of Southampton's city walls. The limestone is an excellent freestone and ashlar capable of being finely shaped into rectilinear blocks.

Bembridge Limestone is composed of numerous shells cemented by crystalline calcite. The freshwater gastropod *Galba* is a common fossil, which, when removed in solution, leaves cavities or vugs. By contrast, Quarr Stone is a highly localized shelly limestone composed of small, non-marine bivalves swept together by storm action. Where the shells dissolve, the resulting moulds remain as voids or are filled with calcite.

Mudstones and Sandstones

Fast-flowing mountain streams erode narrow V-shaped valleys. Erosion of the channels occurs by hydraulic action and corrasion, and the fast-flowing rivers transport boulders, cobbles and pebbles as bedload, particularly during floods. Finer sediment is carried in suspension downstream and deposited when current velocity is minimal in lower reaches of the river. As a lowland river meanders over its floodplain, deposition of silt takes place when water overflows the river banks. Lag gravels and coarser sands accumulate on the point bars on the inside of meander bends, where current is slower.

A delta may develop on a coast where a slow-moving river is heavily charged with sediment and forced to deposit its load as current velocity drops. Fine silt may be transported many miles out to sea. When sea level fluctuations occur on a deltaic coast, repeated fining-up sequences, or cyclothems, of siltstone, shale, sandstone and coal seams are produced.

In contrast to a meandering river, a braided river forms where the current velocity is high and there is an abundant supply of coarse material forming the bedload. A braided river consists of many channels separated by islands of poorly sorted pebbles and cobbles. This often occurs where a river, fed by glacial melt water, loses velocity when it emerges from a mountain tract onto lowland. Conglomerates are formed when large, rounded clasts are deposited as the velocity of the current reduces rapidly.

Turbidity currents occur where sediment is washed out over the continental shelves and descends rapidly down the continental slope through submarine canyons. Poorly sorted, slumped, coarse-grained sediment will form a rock called greywacke, while finer muds and sands will accumulate as a fining-up sequence.

Some Sedimentary Rock Structures

The most common structure found in sandstones is cross-stratification, also known as cross-bedding. This can be observed where the internal stratification within a bed is inclined at an oblique angle to the bedding planes. Aeolian cross-bedding is seen in desert sandstones where the wind shapes the dunes, moving the sand grains up the windward slope and producing steeply inclined foreset beds on the lee side. These dunes exhibit large-scale cross-stratification several metres high.

Deltaic cross-bedding is formed on the delta front as it advances seawards. Ripple bedding results from either wave action in shallow water (symmetrical ripples) or river current action (asymmetrical ripples). It is important to recognize that these forms of cross-stratification are produced by the down-current migration of sediment in river channels and shallow waters. The direction of flow is

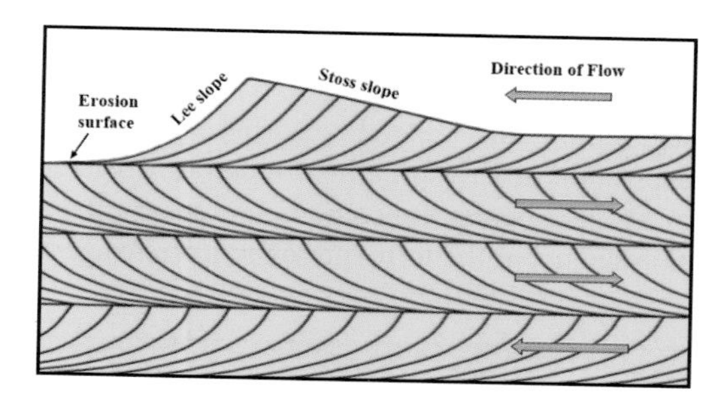

ABOVE: **Cross-stratification or cross-bedding, with asymmetrical ripples (in water) or dunes (wind-blown) migrated in the direction of flow. Each set develops on the eroded surface of older deposits.**

LEFT: **Cross-stratification in a fine-grained sandstone in a beach gulley near Whale Chine. The current was flowing right to left.**

shown by the inclination of the cross-beds and this is of use to geologists in determining the direction of movement of the palaeocurrent. Cross-bedding is best seen in vertical cross-sections at right angles to the bedding planes.

Graded bedding occurs where there is a gradual decrease in the velocity of the current so that the heavier, larger clasts/grains are deposited first, followed by increasingly smaller grains as the current slows. This produces a fining upwards sequence in a bed of sedimentary rock. Graded bedding is also associated with turbidity currents, where deposition of sediments take place when strong currents slow down at the base of submarine slopes.

When rivers meander across floodplains or deltas, they cut channels through the underlying sediments. These erosional river channels can be recognized in section by their concave outline that cross-cuts the underlying beds. The channel infill often contains lag gravels at its base, with a fining-upwards sequence of sediment.

Slump structures are a type of soft sediment that deform when water-saturated material slides down a depositional slope during early stages of consolidation. Convolute bedding forms when complex folding and crumpling occur in clays, silt and fine-grained sands.

Nodules and concretions are formed by the segregation of mineral material soon after deposition. Strictly speaking, nodules are irregular in shape whereas concretions are spherical or ellipsoidal, but the terms are often interchanged.

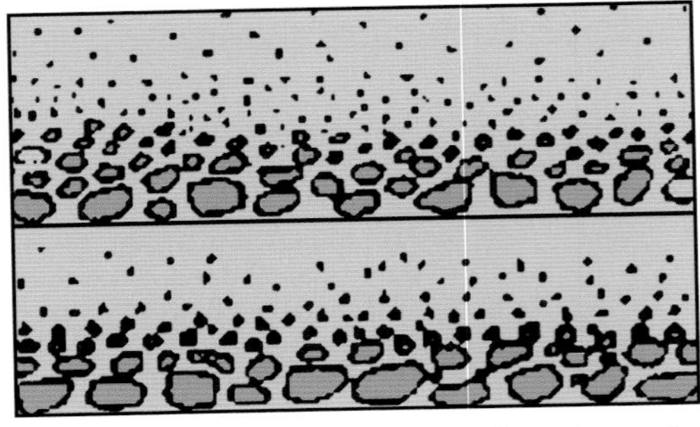

Two units showing graded bedding, fining upwards from lag gravels through sandstone to mudstone.

Concretions consist of carbonate minerals such as calcite and siderite and are found in mudstones and shales; whilst nodules are usually formed of silica in the form of flint and chert.

Palaeogeography

At the beginning of the Lower Cretaceous (140–100 Ma), southern England was situated around 45 degrees north, where it experienced a warm temperate climate. The Wessex Basin (including the Isle of Wight and the Weald) was a subsiding area of fluvial and deltaic deposition that lasted for nearly 30 million years. This basin was separated by the Anglo-Belgian Axis from the marine basin to the north. From time to time the two basins were connected as sea level rose and the waters overflowed through the Bedfordshire Straits.

Much of the rest of Britain was a landmass bounded on the northwest by the Rockall Rift, where sea floor spreading was associated with the opening of the North Atlantic. Worldwide plate tectonic activity, leading to an increase in the volume of mid-oceanic ridges, caused a eustatic rise in sea level during the Cretaceous that was almost as high as the peak reached during Ordovician times. Towards the end of the Lower Cretaceous, the Tethys sea extended northwards, inundating most of southern England and, as sea levels rose, it overflowed the Anglo-Belgian Axis into the northern basin. This marine transgression deposited first the Lower Greensand, then later the Gault Clay and Upper Greensand. The increasing extent and depth of water can be seen by the successive overstepping of older beds as the sea spread farther west. The Upper Greensand oversteps the Gault and older rocks to cap the Haldon Hills in Devon, where it rests directly on Permian strata.

The Upper Cretaceous (100–65 Ma) was a time of high sea level (300m higher than today), when marine transgression extended over most of Britain and northern France. Probably the Scottish Highlands, the Pennines and the Cambrian Mountains stood as islands surrounded by a sea in which fine-grained

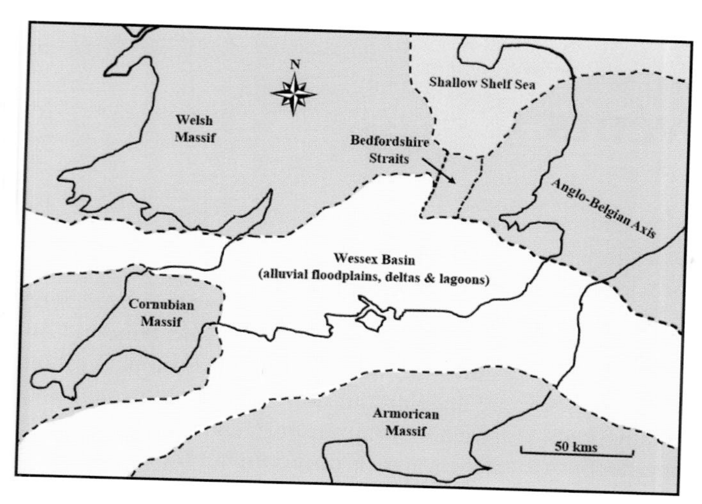

LEFT: **Palaeogeography of the Lower Cretaceous (Wealden) in southern Britain.**

Each of these maps is a general interpretation of the distribution of land and sea for its named period of geological time.

BELOW: **Palaeogeography of the late Eocene in southern Britain.**

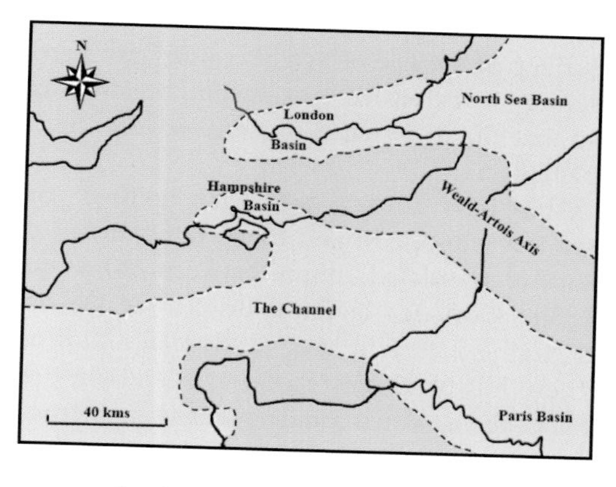

calcareous mud was initially laid down. But as time went on, very little sand and mud was deposited, suggesting that the marginal land areas had been reduced to lowlands. This lack of terrigenous material brought down by rivers means that the chalk sediment is very pure and uniform in composition. Due to subsequent erosion, the chalk outcrop that we see today is only a fraction of that which formerly existed.

The chalk sea withdrew from Britain at the close of the Cretaceous as earth movements uplifted much of northwest Europe. (These were the precursors of the Alpine movements that were to come.) A period of at least 15 million years elapsed before the first Palaeogene sediments were laid down, during which time some 150m of chalk had been eroded in southern Britain.

The palaeogeography of the Palaeogene (66–23 Ma) was dominated by development of the Anglo-Parisian curvette, in which marine deposits of sands and clays were interstratified with fluvial and estuarine sediments. The resulting strata are now preserved in the synclinal structures of the London and Hampshire basins, separated by the Weald-Artois anticline (also to the south lay the Paris basin). Throughout the Eocene and early Oligocene, marine sedimentation continued as the sea repeatedly advanced from the east. However, from time to time, when marine regression occurred, rivers transported sediments from the west, and muds and sands were deposited in an estuarine and lagoonal environment.

While relatively quiet conditions existed in southern Britain, things were very different in northwest Scotland and northern Ireland during the late Palaeocene and early Eocene. Here, intense volcanic activity developed, associated with rifting and sea floor spreading in the North Atlantic. Major igneous intrusions occurred across the Hebrides, including Skye, Mull and Arran. Plateau basalts such as in Antrim (Giant's Causeway) were erupted and dyke swarms issued from the volcanic centres. Long after the vulcanism had died down, during the Miocene the outer ripples of the Alpine orogeny caused uplift, folding and faulting in southern Britain.

CHAPTER 6

Quaternary Events

After the Bouldnor Formation, no younger 'solid' rocks are known in the Isle of Wight. The Miocene period was a time of intense tectonic activity, when the existing rocks were folded and faulted, including the Purbeck-Wight monocline. The asymmetry of many folds suggests that the pressure came from the south as the Alpine orogeny reached its peak. When the earth movements subsided, the uplifted surface of the Isle of Wight would have been subject to extensive erosion and weathering so that the fold structures were reduced to a peneplain.

This early surface was later uplifted (or sea level dropped), and today remnants of it survive as isolated hilltops between 200m and 250m above sea level on Brighstone Down, St Catherine's Hill and St Boniface Down. These hills may have stood as islands in the Pliocene sea, which could have extended across the Isle of Wight, but there is no evidence of any Pliocene marine deposits here, although correlation has been made with a similar planed surface covered by marine sands on the North Downs at Netley Heath in Surrey and Lenham in Kent.

Gravel Deposits

Angular Flint Gravels

These are scattered across the chalk uplands on the Isle of Wight and these are the local equivalent of the Clay-with-Flints that are widespread in southern England. These deposits are clearly shown overlying the chalk on the BGS Special Sheet of the Isle of Wight. They are considered to be the insoluble residue (flints and iron hydroxides) left behind by the gradual removal of chalk in solution by rainwater. It is also possible that some of the flints and clay could have been derived from the overlying Reading Beds and London Clay before they were removed by erosion. The flints are angular and unworn since they obviously have not been transported by water. The age of these gravels is uncertain but they are thought to have been formed during the Pliocene and early Pleistocene.

Plateau Gravels

Composed mainly of rounded and sub-angular flint pebbles, these gravels are found on uplands around 100–120m above sea level. They are the remnants of a once extensive sheet of gravel that spread over the land surface that has since been much dissected by a close network of rivers. These gravels were probably deposited by a large early Pleistocene river that flowed northwards off snow-covered uplands far to the south of the present Isle of Wight.

The Plateau Gravels on St George's Down near Newport are several metres thick and lie on a surface planed across steeply dipping Chalk, Upper Greensand and Gault. On Headon Hill, the gravels have a sandy matrix and in some of the old gravel pits there is evidence of cryoturbation, where freeze-thaw action has produced contorted beds within the partially stratified gravels.

River Gravels

These are typically composed of rounded pebbles set in a sandy matrix that are found on river terraces on the sides of river valleys. A river terrace is a remnant of a former valley floor that stood at a higher level than the present river. When rejuvenation occurred,

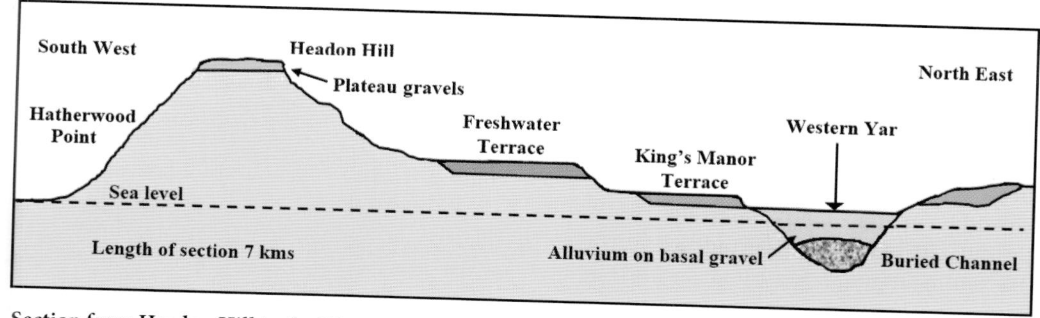

Section from Headon Hill to the Western Yar Valley, showing Quaternary deposits.

due to uplift or a fall in sea level, the river was forced to cut a new valley within its former floodplain to adjust to a lower base level. Most of the rivers on the Isle of Wight have some river terraces, some of which are at different levels where rejuvenation has occurred more than once. In the cross-section from Alum Bay to the Western Yar valley, you can see that there are two terrace levels above the present alluvial floodplain, and at the top of Headon Hill there is a capping of Plateau Gravels.

Other examples of terraces can be seen in the Medina River valley near Godshill, where the Bridge Farm terrace is well preserved, and in the Eastern Yar valley, where Hale Common [SZ 550840] stands on an extensive spread of terrace gravels. The higher terraces in the upper reaches of the rivers probably date from the late Pleistocene.

Marine Gravels

At around 35–60m above sea level on the northeast coast of the island, you will find marine gravels. They are mapped as the Wootton Gravel Complex (on the BGS Special Sheet of the Isle of Wight), consisting of well-rounded pebbles and sand that have been reworked by wave action when the sea stood at a higher level during the Ipswichian interglacial.

Probably the best evidence of a higher sea level in the past is provided by the Bembridge Raised Beach. This rests on a platform cut into the Bembridge Marls, with the base of the former cliff about 15m above the present sea level. The rock platform and the cliff are buried beneath several metres of flint gravel, sand and silt, all of which are covered by a layer of brickearth that is banked against the old cliff line.

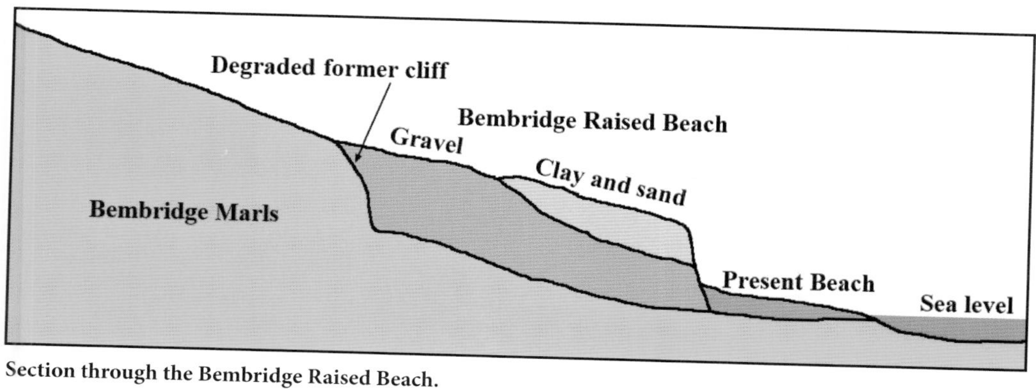

Section through the Bembridge Raised Beach.

Periglacial Deposits

During the last glacial episode, known as the Devensian, the ice sheets did not reach as far south as the Isle of Wight, but nevertheless periglacial conditions prevailed. Winds blowing across cold, sparsely vegetated unconsolidated outcrops would collect loose dry sediment and deposit it as an unstratified fine-grained loamy mixture of sand and clay referred to as brickearth or loess. This wind-blown superficial deposit is a fertile type of soil that today is blown out of the cold Gobi Desert to accumulate in northern China. Brickearth can be seen covering part of the old sea cliff behind the Bembridge Raised Beach.

Another type of periglacial deposit is one that geologists call head (sometimes referred to as rubble drift in older literature). This is produced by the downslope movement of weathered surface material under repeated freeze-thaw conditions by a process known as solifluction. The BGS Special Sheet shows several head deposits located at the base the northern slope of the Central Downs between Calbourne and Carisbrooke. There is also a large isolated deposit near Rookley [SZ 508840], on the western slopes of the Medina valley.

The chalky rubble and flints that accumulate on the lower slopes and floors of dry valleys is known as Coombe Rock. It was formed by alternate freezing and thawing of saturated chalk and often moves downslope due to solifluction; it is thus a form of head restricted to chalk country.

Holocene deposits are those laid down after the end of the Pleistocene, some 12,000 years ago. They include alluvium, peat and sand dunes. Alluvium is the accumulation of fine sediment on flood plains deposited when the river is in spate. Brading marshes, for example, occupy a large area of alluvium on the flood plain of the River East Yar. Freshwater peat formed from decomposing swamp vegetation, can be seen in the upper Medina valley and estuarine peat develops in the saltwater marshes of river estuaries. Wind-blown sand often accumulates behind beaches on spits when onshore winds are strong and persistent. The Duver across Bembridge Harbour and the Newtown Harbour spit both demonstrate sand dune topography.

Geologists consider that an eastward-flowing Solent river system was well established by mid-Pleistocene times, when the Isle of Wight was still part of the mainland. When sea level

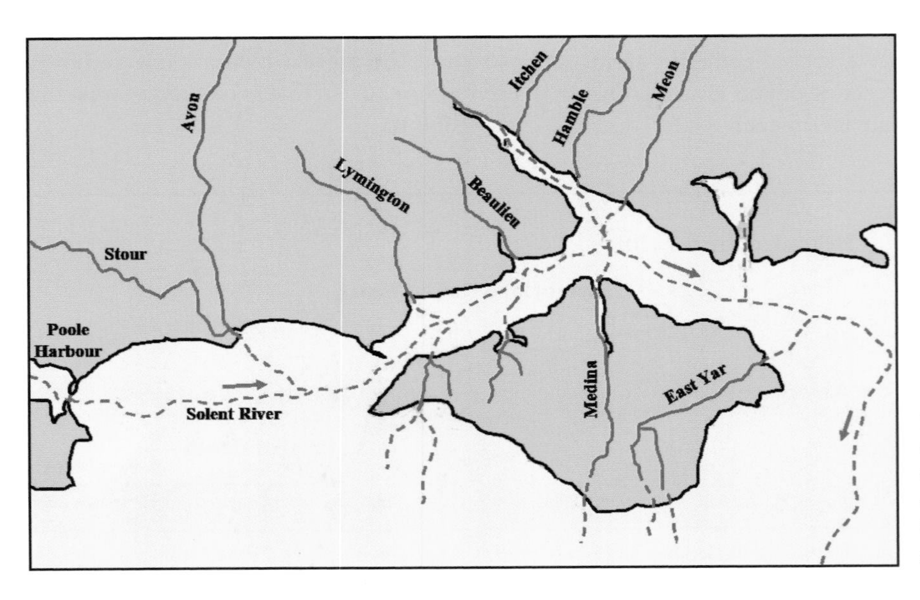

The Solent river system that was active during the Devensian glacial stage.

CHAPTER 7

Sculpting the Landscape

The landscape of the Isle of Wight owes much to the work of rivers, which erode the uplands and subsequently transport and deposit sediment on the lowlands. In its upper course, a river will actively erode its channel through hydraulic action when the current is strong and capable of removing rock fragments, cobbles and boulders. The process of corrasion also wears away the sides and floor of the channel by the scouring action of the water armed with pebbles and coarse sediment. In chalk and limestone areas, calcium carbonate is removed in solution thus enlarging the channel. As the velocity of the current decreases downstream, the river begins to meander over its floodplain, depositing silt and sand on the surrounding area when it is in spate. This sediment is known as alluvium, a superficial deposit that is mapped by the BGS along the lower stretches of the Western Yar, Newtown River, Medina and Eastern Yar.

The Early River System

As we saw in the previous chapter, the river system of the Isle of Wight was part of the much larger Solent river system during the Devensian, when sea level was considerably lower than now. Rivers would have been much longer and would have been initiated in Pliocene times on an extensive chalk surface that was gently sloping to the north. These rivers are called consequent streams by geomorphologists since they are consequent on the slope of the initial surface. They become powerful rivers fed by melt water and incise their channels into the underlying strata as the chalk cover is largely removed. Thus, today you will notice that these rivers have cut across the east–west strike of the rock outcrops and sawn through the chalk monoclinal ridge to create gaps at Freshwater Bay (Western Yar), Newport (Medina) and Brading (Eastern Yar).

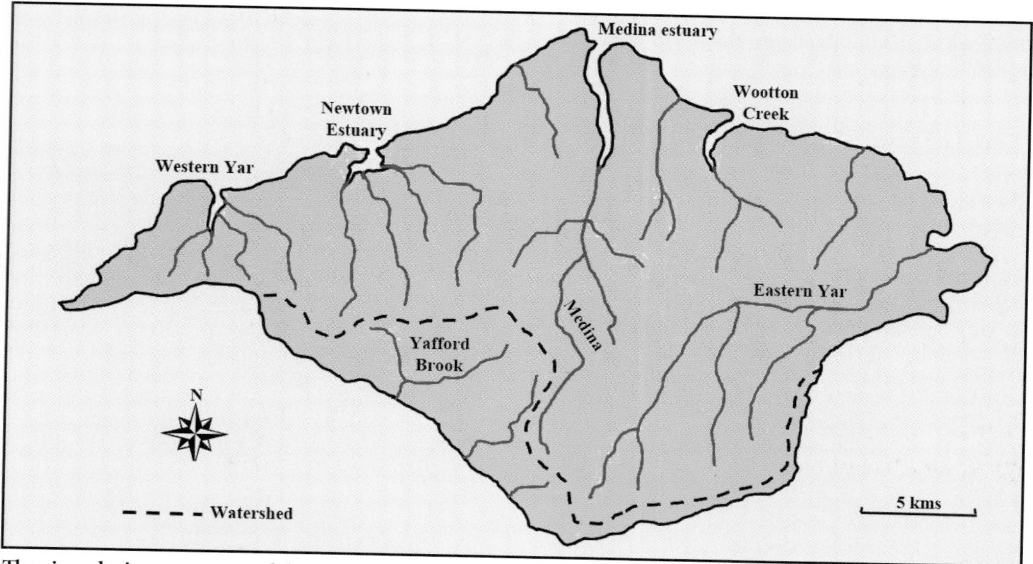

The river drainage pattern of the Isle of Wight.

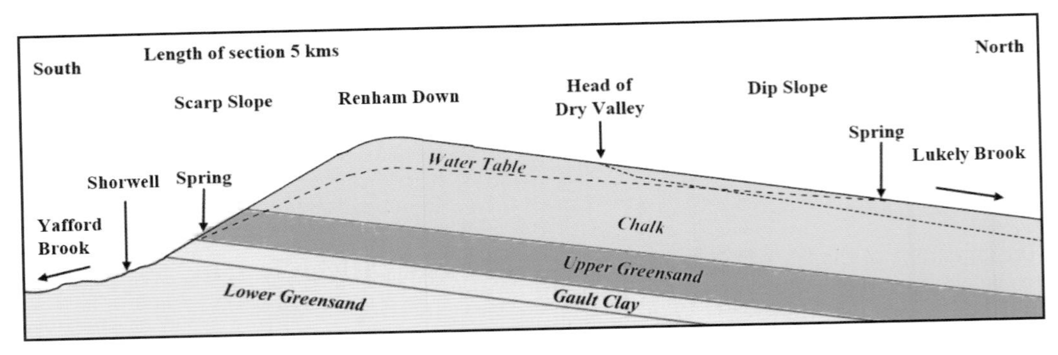

Section through the chalk escarpment and dip slope near Shorwell.

There are also, however, some tributary streams that developed later and now follow the strike of some of the softer clay outcrops. These are termed subsequent streams; for example, the Thorley Brook is a subsequent tributary of the Western Yar, working its way along the outcrop of the Bembridge Marls. The Eastern Yar makes a sharp turn just north of Sandown and follows an east–west course through Alverstone as far as Horringford [SZ 543853], following the outcrop of the Ferruginous Sands. Before its headwaters were removed by marine erosion, the main river flowed northwards and the present east–west stretch would have been a subsequent tributary stream.

Rivers Flowing North

A glance at the drainage map of the Isle of Wight will show you that around 80 per cent of the island is drained by northward-flowing rivers. This is not surprising since the regional dip of the strata on the island is to the north. These rivers are separated from the southward-flowing rivers by a watershed that runs through the crest of the southern chalk downs then turns north before following the central chalk outcrop as far west as the Needles. One reason the south-flowing streams are so short is that their headstreams have been removed by coastal recession along the south coast. The Western Yar, in particular, has been beheaded and is now a small 'misfit' stream flowing through a wide valley cut by its much larger predecessor. Gravel terraces on the sides of the valley bear witness to the size of the former river that flowed at a higher level.

The road across the chalk downland with the island's southwestern coast fading into the distance.

However, some longer north-flowing rivers have also suffered modification in a process known as river capture. This happens when one stream cuts back by headward erosion through soft rocks at a lower level than a stream on the other side of a drainage divide, thus undermining it and diverting the waters into its own system. For example, look at the Ordnance Survey 1:50,000 Sheet 196 and you will see that near to Bohemia Corner [SZ 523836], south of Merstone, the Eastern Yar has captured some of the former headstreams of the Medina while beheading the Merstone stream, which is now a much shorter misfit in a wide, open valley.

There is another interesting drainage modification in the area around Shorwell [SZ 458830], where the small, south-flowing Yafford Brook has cut back through the soft Gault Clay and Upper Greensand and now has its source in the Lower Chalk. As a result, the headwaters of the Lukely Brook, a tributary of the Medina, have now become dry, as water is diverted underground below the watershed into the Yafford Brook. Furthermore, as scarp recession continues due to weathering and headward erosion, the level of the scarp foot springs will fall, thus lowering the water table and increasing the size of the dry valleys on the dip slope. The dry valleys were probably excavated in late Pleistocene times when the periglacial chalk landscape began to thaw out and the water table was much higher than today.

The Chines

Now let us consider the short streams that flow south from the main watershed. Some of these streams drain the steep chalk scarp of the southern downland between Dunnose and St Catherine's Point, while others are initiated on the Wealden and Lower Greensand lowlands around the southwest coast. Most have steep gradients cutting narrow ravines through the cliffed coastline and are generally known as chines in the Isle of Wight. Downcutting is intensified and gradients steepened as coastal erosion and groundwater seepage constantly undercuts the cliffs and shortens the streams.

The development of Blackgang Chine since the nineteenth century provides a good illustration of the way in which groundwater undermines cliffs and modies the mouth of a chine. Drawings made by W.H. Fitton in the 1840s show a steep, V-shaped cross profile with marginal ledges of hard ferruginous bands within the Lower Greensand. The stream cascaded over the ledges, creating a waterfall down to the shore. Since then the upper cliff has been slowly cut back by over 100m due to seepage at its base where groundwater emerges, creating outwash and leading to the undermining and collapse of the sandstone cliffs and the destruction of the chine. Fans of sandy mud are swept across the undercliff terrace and over the lower cliff, eventually to be removed by the waves.

continued on p.73

The Isle of Wight's interior lowland, formed on the weaker Ferruginous Sands and clays, looking northwards from Appuldurcombe Down with the village of Godshill in the foreground.

Itinerary 13: Along the Chalk Downs from Brading to Freshwater Bay

Walking distance: two sections, 8km and 14km

This itinerary can be followed in two stages to fit time available and the visitos' fintness. The eastern section is suitable for cycling since it follows the minor road along the crest of the chalk downs.

The start is at Yarbridge [SZ 605864], where the Eastern Yar breaches the chalk ridge to create the Brading gap. Climb the lane up to the viewpoint on Brading Down. Looking northeast from here you will see the alluvial marshlands behind Bembridge Harbour and the Brading Marshes Nature Reserve. To the east, beyond the Brading gap, the chalk extends across Bembridge Down, with its distinctive Iron Age fort, to the coast at Culver Cliff. To the south, the narrow outcrop of the Chalk quickly gives way to the Upper Greensand scarp and the undulating country of Ferruginous Sands around Adgestone, which overlooks the lowland in the eroded Sandown anticline. Further south, on a clear day, you should be able to identify the chalk salients of St Martin's, Appuldurcombe and St Catherine's Down marking the edge of the southern plateau. To the southwest is the extensive lowland floored by the Ferruginous Sands and drained by the headwaters of the Eastern Yar.

Continue along the road over Ashey Down to Mersley Down (125m), for another excellent view. The chalk outcrop is only about 500m wide since the strata dip steeply to the north; thus the scarp and dip slope have roughly equal gradients and so this is a classic hog's-back landform. The view northwards is across undulating lowlands formed of soft sands and clays of Hamstead Beds and Bembridge Marls. On the coast, Ryde ferry port is clearly visible, then to the west is Wooton Creek, the Medina estuary and the yachting centre of Cowes. The road continues to the Hare and Hounds at Downend, where drinks can be found.

The second section begins at Carisbrooke, with its Norman castle where Charles I was held prisoner in 1648 before being taken to London for his execution. From Carisbrooke, follow the Tennyson Trail southwest to Bowcombe Down, which is capped by a spread of angular flint gravel. We are now starting to cross the central downs, which are about 6km wide at this point due to the gentle northerly dip of the chalk. This is in sharp contrast to the narrow chalk outcrops east of Newport and west of Freshwater.

Continue along the trail to Gallibury Hump, which rises to just over 200m above sea level. This eminence allows an excellent view northwards across the chalk dip slope, giving you a good idea of the dissected topography due to the numerous dry valleys opening out to the north and east. There is also evidence of prehistoric settlement with Neolithic burial mounds (marked on the O.S. map as tumuli: *see* Chapter 9).

The Tennyson Trail continues through Brighstone Forest, occupying the summit of Brighstone Down (214m), and then emerges along the crest of the chalk escarpment. When you reach the car park where Lynch Lane crosses the trail, notice the small subsequent valley with its stream flowing out from a scarp foot spring known as Buddlehole Spring [SZ 423843] at the junction of the Upper Greensand and the Gault Clay. Walking towards Mottistone Down, the view to the south is over the Gault Clay vale at the foot of the scarp. Beyond is a ridge of high ground that is formed of Lower Greensand (Ferruginous Sand, Sandrock and Carstone), which has a low scarp overlooking the lowland developed on the Wealden Beds. Note that in this area the beds dip gently NNE and the strike of the parallel outcrops is WNW–ESE.

The trail continues to the Chessell gap, where the chalk downs are cut by one of the headstreams of the Caul Bourne flowing north into the Newtown estuary. As you walk westwards towards Compton Down, the chalk outcrop narrows as the northerly dip increases and, by the time Afton Down is reached, the chalk has once more assumed the characteristics of a hog's-back landform. Finally, at Freshwater, the chalk ridge is breached by the sea but continues west over Tennyson Down to the Needles.

The central downs route from Brading to Freshwater.

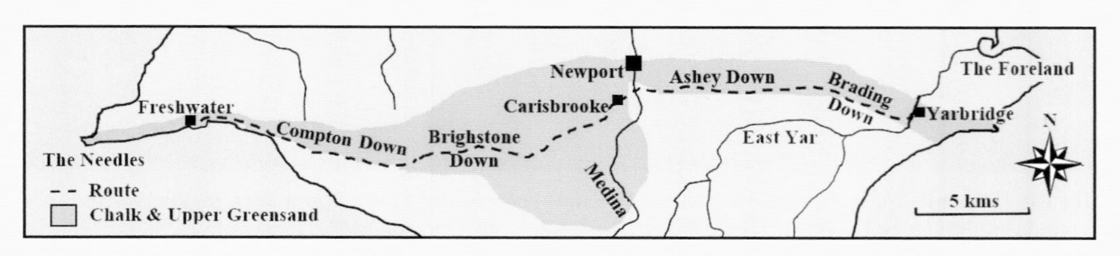

Itinerary 14: A Walk on the Southern Downland

Walking distance: 14km

This circular walk begins and ends in Ventnor and is designed to show some of the topographical features of the chalk plateau in the south of the island.

First, climb the steep zig-zag route up through the town and join the track leading over Rew Down to the Ventnor golf course. Continue northwards on the Stenbury Trail, noting the three tumuli on the west side of the path, evidence of Bronze Age burials. The path follows the chalk ridge, capped with clay with flints, over Stenbury Down towards the communications mast at SZ 538793. The ridge forms the interfluve between two north-flowing headstreams of the Eastern Yar; namely the Whitwell and Wroxall streams. Both valleys are cut into the Sandrock, Gault Clay and Upper Greensand, which have been exposed by continuous fluvial erosion. Since the chalk dips gently southwards, the track climbs the dip slope up to the summit of Appuldurcombe Down, which is marked by a second mast at 226m above sea level.

Just before the summit, take the path to the right, which leads down the chalk escarpment to a bench feature in the cherty sandstones of the Upper Greensand that ends in the scarped edge of Gat Cliff. Beyond the cliff lies an apron of Gault Clay that produces landslipped hummocky, broken ground. The Worsley Trail runs below Gad Cliff and this should be followed to the gates of Appuldurcombe Park [SZ 540807]. Then take the footpath southeastwards to Wroxall village, where refreshments are available at the Star Inn and other establishments.

From Wroxall church, a lane leads past the cemetery to where the trail heads northeast to Cook's Castle [SZ 558805] on the well-defined Upper Greensand bench below the escarpment of St Martin's Down. This is the equivalent geological structure to that seen at Gat Cliff. Continue southeastwards to the summit of Shanklin Down at 235m above sea level, from where there are good views towards the coast at Shanklin and westwards across the Wroxall valley to Appuldurcombe Down. Return along the chalk ridge of Luccombe Down and then on to St Boniface Down before descending into Ventnor.

During this second part of the walk, you should have plenty of opportunity to observe the sinuous morphology of the chalk scarp. The chalk plateau is extremely dissected by the headstreams of the Eastern Yar. Each re-entrant valley is separated from the adjacent one by a salient or spur of high land; for example, the Greatwood Copse valley below Shanklin Down is separated from the Luccombe valley by a well-defined salient, and this is emphasized by the way that the A3055 road curves over the spur and back into the head of the re-entrant valleys. This indented pattern around the margin of the chalk outcrop is typical of a mature dissected plateau, which is seen particularly well in this southeastern corner of the Isle of Wight.

Finally, you will notice that the descent into Ventnor is across a salient overlooking the St Boniface dry valley [SZ 563782]. This is a dip slope valley draining south, and there is a similar one on the south side of Week Down [SZ 545775], near the start of the Stenbury Trail. These dry valleys would have been cut by streams flowing southwards when the chalk extended far beyond the present coastline before it was removed by marine erosion.

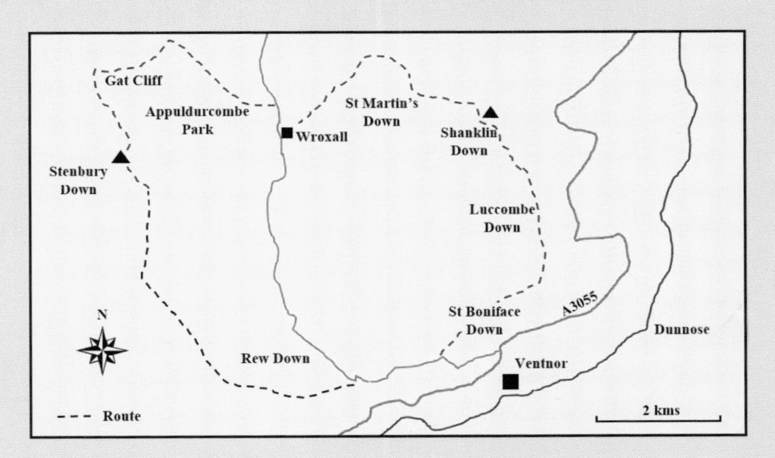

The circular route from Ventnor that loops around around the southern chalk downs.

Itinerary 15: The Yar River Trail, Niton to Bembridge Harbour

Walking distance: 24km

Niton to Godshill

Let us begin at the beginning: the longest river on the Isle of Wight is the Eastern Yar, with a length of 20km, and its source is a spring on the western outskirts of the village of Niton at SZ 503766. The spring issues at the junction of the permeable Upper Greensand and the impermeable Gault Clay near the head of a valley on the eastern slopes of St Catherine's Hill.

The trail follows Ashknowle Lane out of Niton in a northeasterly direction towards Whitwell. It runs along the east side of the Yar valley at the foot of the scarp below the outcrop of the Chert Beds in the Upper Greensand.

Continue through Whitwell, which stands on a broad embayment of Gault Clay in a tributary valley of the Yar, and then turn left on the track alongside Stockbridge House, crossing to the west side of the Yar at the footbridge. As you approach Southford, notice that the valley is now developed on the Sandrock Beds that have a wide outcrop extending north to Roud hamlet.

The trail now follows the river past Ford Farm and a sewage works, where there are several treatment ponds utilizing river water. Above Roud there is a shortcut riverside path that brings you to where Beacon Alley crosses the river. [SZ 519813] This section of the river is cut into a gravelly terrace, a former floodplain, marked on the BGS map as the Bridge Farm terrace. The trail then continues north and crosses the River Yar near Bridgecourt Farm and heads east to Godshill.

Here there are excellent views to the south of the Sandrock escarpment, which rises behind Sainham Farm [SZ 528810] and beyond to Gat Cliff, where the cherty sandstone of the Upper Greensand outcrops. All Saints Church stands on a spur of sandstone within the Ferruginous Sands Formation, which stretches in a broad band across the island from Shanklin to Chale Bay. The medieval church is built of local stone and stands surrounded by thatched cottages in a picturesque village setting. Godshill may be a suitable place to stop for rest and refreshment before embarking on the next stage of your journey.

The Eastern Yar Floodplain

From the church, cross the main road and take the path leading northeast to Moor Farm and onward to Great Budbridge Manor, where you will again cross the Yar. A short diversion to the west of about 500m along the path that crosses the disused railway line will bring you to a point [SZ 523836] where you can view the large bend in the river near to Bohemia Corner. This bend is significant (to geomorphologists) because it is what is known as an 'elbow of capture', where the Eastern Yar has, at an earlier time, cut back and captured or diverted one of the headwaters of

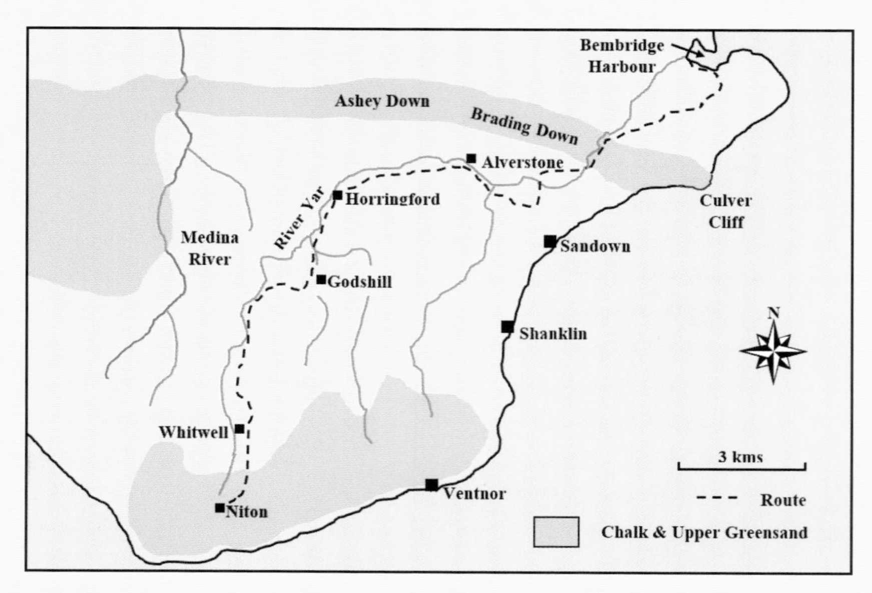

Map of the Yar Valley Trail from Niton to Bembridge Harbour.

the Medina. In other words, the Upper Yar that we have been following was previously a headstream of the Medina River that flowed north through Merstone and Blackwater and beyond to Newport.

Retrace your steps to Great Budbridge Manor and follow the trail to Horringford. If you wish to follow the river more closely, then it is possible to join the cycle track along the disused railway line that crosses the trail just north of Little Budbridge Farm. This section of the river flows over a floodplain composed largely of freshwater peat, which was formed over thousands of years from the accumulated rotting vegetation on the swampy terrain. The numerous ponds alongside the river probably mark the site of former peat diggings and some are now used as fishing lakes. On the east side of the river, above 25m, is Hale Common, a gravel terrace deposit with a rich brickearth cover that is the site of glasshouses and polythene tunnels used by commercial nurseries.

The next section of the trail continues through Landbridge to Alverstone as the river flows across an alluvial floodplain bordered by bluffs of Ferruginous Sands on either side. The chalk hog's back of Mersey Down and Ashey Down form the skyline to the north. Alverstone Mead Nature Reserve encompasses the marshy floodplain where the wetlands provide a haven for wintering wildfowl such as teal and snipe, plus herons, barn owls and kingfishers.

The trail continues to the outskirts of Sandown before turning north to Adgestone, where vines are cultivated on the well-drained south-facing slopes on the Ferruginous Sands in the lee of Brading Down. A Roman villa with well-preserved remains of mosaic floors is located immediately south of the trail at SZ 599862. Descend into Yarbridge, where the river flows through the Yar Gap between Brading Down and Bembridge Down. The pathway follows the line of the railway for 600m then turns eastwards across Brading Marshes, where the river meanders across a wide floodplain. The RSPB nature reserve is based in the marshland, which attracts a range of birds including marsh harriers, egrets, and buzzards.

After crossing the floodplain, the trail runs around the base of Centurion's Hill and then eastwards to Bembridge Windmill, which was built in the early 1700s and is now owned by the National Trust. The windmill stands on a spur of higher ground formed of the Bembridge Marls, and from here you can view the tidal inlets running down to Bembridge Harbour and the course of the River Yar as it crosses the lower marshes. Here the river flows through an artificially straightened channel above the level of the surrounding alluvial plain, which can be subject to tidal river flooding. Leave the windmill and walk down to Bembridge Harbour in the tidal estuary, which is protected from the sea by paired sand spits.

Itinerary 16: A Walk from Blackgang to Newport

Walking distance: 20km

Note: this route does not pass through any sizeable villages, so it is recommended that you carry refreshments with you.

This walk begins at the carpark in Blackgang [SZ 490765] on the A3066. The view southwards from the car park is across the Gore Cliff landslip, where broken masses of Chalk and Upper Greensand have slipped down over the Gault Clay. To the west, you can see the mudslides of Blackgang Chine and Chale Bay.

The path from the car park climbs over the Upper Greensand bench up the chalk slopes of St Catherine's Hill, 236m above sea level. Near to the summit is the octagonal tower of St Catherine's Oratory, which served as a medieval lighthouse and is built of local sandstone. There are extensive views to the west along the coast of Chale Bay and Brighstone Bay and northward over the Lower Greensand lowlands, drained by the Upper Medina, to Chillerton Down and surrounding chalk hills.

The walk continues past Tolt Rocks, an outcrop of the Chert Beds, along the narrow Upper Greensand ridge to Hoy's Monument. On either side of the ridge are landslipped slopes that developed under periglacial conditions, when the sandstone broke up and moved downslope over the water lubricated Gault Clay.

Follow the small valley of the Rew, a headstream of the Medina, cut into the Sandrock Beds, then turn north on the footpath to Upper Appleford Farm, descending from the Sandrock bench onto the

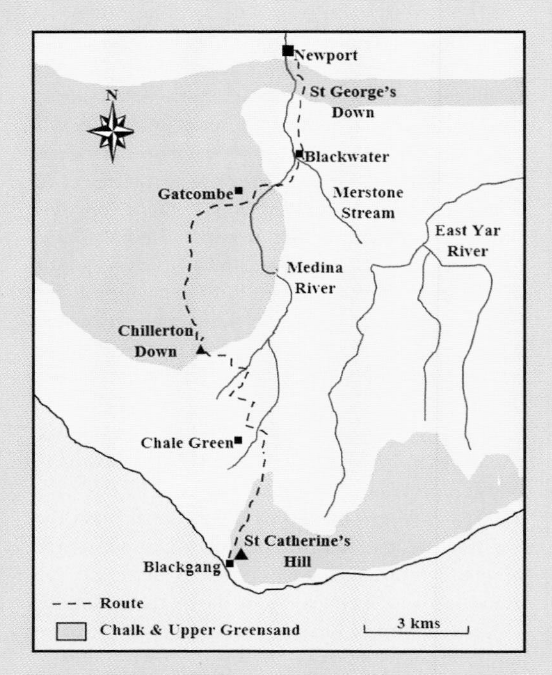

Map of the walk from Blackgang to Newport via the chalk hill of Chillerton Down.

underlying Ferruginous Sands. The footpath joins a lane just beyond the farm; turn right and after 400m turn left onto a footpath that crosses the Medina adjacent to a sewage works [SZ 491807]. Here the river is flowing over a marshy floodplain that widens downstream in an area known as The Wilderness, where deposits of freshwater peat form ill-drained scrubland containing tussock sedge and willow. The marsh is designated as a SSSI since it is the only example of surviving, freshwater, peatland vegetation on the island.

After crossing the river, take the path across the Ferruginous Sands to Beckfield Cross [SZ 482810], then follow the lane northwards to Billingham Manor, which is overlooked by the south-facing scarp of the Sandrock Beds. About 200m to the north of the manor, turn right onto the footpath leading up to the base of the Upper Greensand scarp of Berry Hill. Join the lane at SZ485816 and continue northwest to the Chillerton road, which bears north through the gap in the Upper Greensand scarp, then take the track to the left at SZ 482829 that leads up the chalk escarpment of Chillerton Down. On the way up the track there are several old quarries in the Lower Chalk where the thin rendzina soils are exposed above the rock section.

When you reach the summit near the TV mast, 167m above sea level, you will notice that the plateau surface running northwards is remarkably level at around 160m OD (Ordnance Datum) and this is because it is a remnant of an ancient land surface that was cut by the Pliocene sea when the chalk extended over most of the southern half of the island.

The surface of the chalk is covered with angular flint gravels mapped as clay with flints by the BGS. The flints are derived from the solution weathering of the chalk so they represent an insoluble residual deposit resting directly on the chalk surface. The earthworks of an unfinished Iron Age fort can be seen on the east side Chillerton Down.

The Lower Medina

Continue northwards to Dukem Copse and then turn sharp right along the path that leads west down a steep-sided dry valley to Newbarn Farm. From here, the track runs over the Upper Greensand bench before descending to the Gault Clay near Sheat Manor. When you reach the lane, turn left and walk about 600m, then turn right onto the footpath that crosses the wooded Medina valley and leads to Champion Farm [SZ 502850].

The Medina has a graded long profile that begins around 60m above sea level near Chale Green and descends to 25m just north of Champion Farm. Here there is a steepening of the gradient as the river cuts its channel between the remnants of a low terrace that can be traced northwards through the Medina Gap.

Follow the riverside path to Blackwater, at the confluence of the Merstone tributary with the Medina. There are several flooded former sand and gravel workings on the terrace deposits of the Merstone valley. The path continues as the Stenbury Trail down to Shide at the entrance to the Medina Gap through the narrow chalk ridge.

St George's Down lies on the eastern slopes of the Medina valley, about 1km from Blackwater. It is an area of Plateau Gravels resting on the planed surface of Chalk, Upper Greensand, Gault Clay and Ferruginous Sandstone at an altitude of around 100m. The coarse gravels are derived from the ancestral Medina that flowed north to join the Solent river in the late Pleistocene. Today the sand and gravel are extracted commercially, but the disused and flooded pits provide excellent sanctuaries for wildlife.

The Island's Soil Cover

Soils are largely derived from the underlying rock and generally consist of weathered rock particles mixed with organic material. Climate plays an important part in soil development since it influences the physical, chemical and biological processes that operate within the soil. On the Isle of Wight there are several different major soil types depending on the local geology.

Chalky Soils

The Chalk downlands are covered with a thin calcareous soil up to 15cm in depth known as rendzina, which consists of a dark humus layer that grades through chalk fragments down to bedrock. The soil is often dry and crumbly since it is well drained due to the high permeability of the chalk. This shallow, nutrient-poor soil supports plants that favour alkaline conditions such as calcareous grassland (short turf) and lime-loving wild flowers within a delicate, specialized ecosystem.

Traditionally, the chalk downlands have been grazed by sheep since the soils are too shallow for arable cultivation. However, on the downland plateaux, where there are superficial deposits of clay with flints, the soils are deeper and richer in nutrients and suitable for arable farming, particularly cereal production. Beech trees, native to southern England, flourish on well-drained calcareous soils, often colonizing the steeper scarp slopes of the downlands.

The soils developed on the Lower Cretaceous and Palaeogene siliceous clastic sediments are composed of various proportions of silt, clay and sand as shown on the accompanying triangular graph.

Clay Soils

The soils of the Weald Clay, as the name suggests, are rich in clay particles. They usually contain around 70 per cent clay particles, 20 per cent silt and 10 per cent sand, which produces a rather heavy soil that is rich in nutrients but tends to get waterlogged in winter due to the impermeability of the close-packed clay particles. However, clay soils do provide good pasture for livestock. By contrast, the Lower

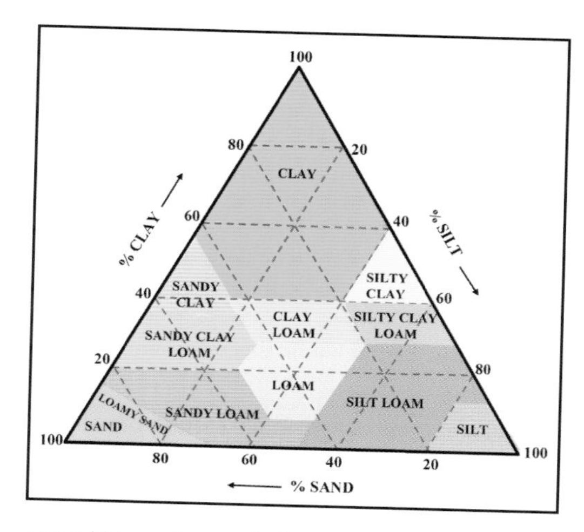

ABOVE: **Triangular graph showing the percentages of sand, silt and clay particles in soils.**

BELOW: **Rendzina soil profile on chalk downlands, showing the dark humus layer above the weathered, fragmented and frost-shattered chalk horizon that grades down toward intact bedrock.**

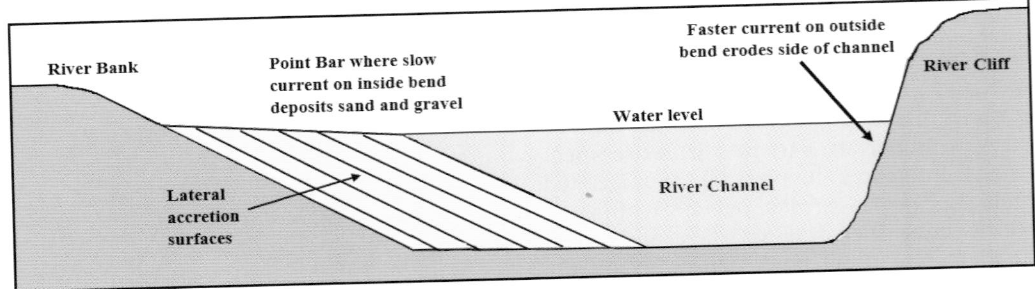

Formation of a point bar on the inside of a river's meander bend.

Greensand tends to produce sandy loams that have up to 60 per cent silica sand, 20 per cent clay and 20 per cent silt. These soils are suitable for arable farming, although grassland pasture is still significant.

In the north of the island, the sands and clays of the Palaeogene strata form rolling undulating country, which supports much surface drainage. This is reflected in the character of the soil cover, which is essentially a clay loam with around 40 per cent clay, 30 per cent sand and 30 per cent silt composition. These nutrient-rich soils are much used for mixed farming in a rural landscape with small fields and hedgerows and numerous copses of deciduous woodland.

Brickearth and Alluvial Soils

Brickearth originated as fine, wind-blown sediment (loess) under dry periglacial conditions and is often preserved above river terrace gravels. It occurs as a homogeneous loam of roughly equal amounts of sand and silt with a small quantity of clay. Its high nutrient content makes it suitable for intensive cultivation, such as market gardening.

Alluvial soils are found along the lower stretches of all the rivers on the Isle of Wight. They accumulate as floodplain sediment when the river velocity is reduced and fine suspended grains are deposited. They contain a mixture of sand, silt and clay, often producing a silt loam containing up to 60 per cent silt particles. These soils are rich in mineral nutrients with a high moisture content and form excellent pastureland. Unfortunately, they are prone to flooding alongside the rivers, hence the term 'water meadow's.

Floodplains, Terraces and Deltas

It is important to understand how rivers deposit sediment in their lower courses because this gives an insight into how some sedimentary rocks are formed. Many of the Lower Cretaceous sediments, such as the Wealden Beds, were laid down by rivers meandering over wide floodplains. When a river reaches its lowland or mature stage, the velocity of its current slows and its suspended load of silt and sand is deposited either as overbank sediment in time of flood or as point bar deposition in migrating meanders. A point bar is where deposition occurs on the inside of a meander bend where the current is weak, forming a series of lateral accretion surfaces marking successive deposits. At the same time the outside bend is being eroded by the faster current, often leading to the formation of a river cliff.

When a river overspills its banks (or levees), silt is spread over the floodplain, where the finest sediment is deposited in static standing water. The annual Nile floods have for centuries covered the surrounding floodplain with fertile soil in which to grow high-yielding crops.

View from St Martin's Down looking northeast towards Shanklin, Sandown and Culver Cliff.

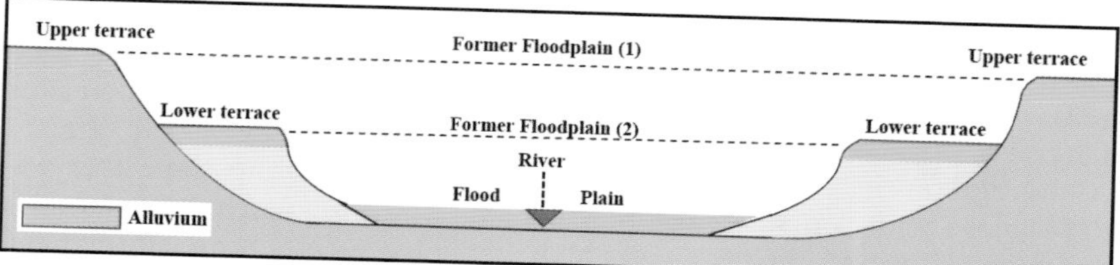

LEFT: **Cross-section of a rejuvenated river valley, that has developed river terraces.**

BELOW: **Structure of a prograding delta.**

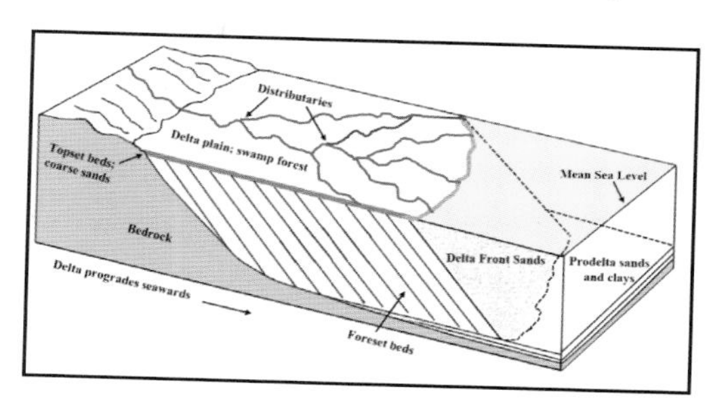

A mature river meandering across a floodplain may appear to be a permanent feature of the landscape but this is often not the case. Meanders will gradually migrate downstream; as the sinuosity of the meanders increases over time, the river in flood will tend to break through the meander neck and develop a new meander further downstream. This process of channel switching is well seen on the Mississippi and it can also take place on delta plains where distributary streams short circuit and switch position under flood conditions. Some rivers that are heavily charged with coarse sand and gravel in their lower reaches will deposit this material to form islands within the main channel, thus dividing the river into several smaller channels and creating a braided river system.

Formation of Terraces

A river grades its course to the lowest base level, which is usually sea level but may be the level of a lake into which it is flowing. Changes in base level occur over a long time-span but they have considerable influence over the course of a river. When the land rises due to earth movements or sea level falls during a glacial period, the river is forced to regrade its long profile to the new lower sea level. This energizes the river flow as it actively erodes a deeper channel, leaving its former floodplain as a river terrace above the newly cut valley. This process of renewed downcutting is known as rejuvenation and the headward limit of the regraded channel is referred to as a knick point. When a river is rejuvenated it will create a valley and floodplain within the old valley, which will be bordered by remnants of the old floodplain or terraces.

But what happens if sea level rises? The river estuaries are flooded and sediment infills the former channel so that the long profile of the river is adjusted to the higher base level. There are numerous examples of river terraces, drowned estuaries and buried channels on the northern coastline of the Isle of Wight (*see* Chapter 6).

Formation of Deltas

When a river loaded with sediment reaches the sea and its velocity is greatly reduced, deposition will occur in the estuary and continue out to sea. Where there are strong tidal currents, they may scour the estuary and prevent it from silting up, but where there is little tidal movement, sediment will build up into a delta plain and cause the river to divide into smaller channels called distributaries.

Deltas grow seawards by a process called progradation, which in effect means that as new sediment is brought down through the distributaries it is deposited on the leading edge of the delta as foreset beds. These are inclined in the direction of the current and, when seen in rock sections, are known as cross-stratified beds or simply cross-bedding. Deltas also show a coarsening upward succession from the finest subaqueous muds of the prodelta through the foreset beds to the coarser sands of the delta plain (topset beds). In other words, the coarser sands are deposited first as topset beds, then the medium grained sands as foreset beds and finally the finest silts and clay particles are carried out beyond the delta front.

Chalk and Water

Chalk is a common rock in eastern and southern England, including the Isle of Wight. Its surface topography is one of rolling hills and dry valleys with a distinct lack of surface drainage. However, below the surface of the chalk, water flows through the interconnected pore spaces between the grains of the rock, which is very permeable. Water is also stored in these spaces due to the high porosity of the rock. The properties of permeability and porosity make chalk an ideal aquifer or water-bearing rock.

The London Basin is an example of a confined aquifer since the chalk is sandwiched between the impermeable Gault Clay below and the impermeable London Clay above. Rainfall on the crest of the Chiltern Hills and the North Downs continually replenishes water within the chalk aquifer. In the 1850s, the water table beneath London was high enough (22m below OD) to create sufficient natural hydraulic pressure to operate the fountains in Trafalgar Square, but subsequently industrial and domestic demand for water lowered the water table drastically. In recent decades, industrial decentralization and licensing controls have reversed the situation and raised the level considerably.

The surface of the chalk downs is often deeply dissected by dry valleys that are tributary to larger rivers. These valleys have obviously been cut at some previous time when surface water was available. It has been suggested that during post-Devensian periglacial times, surface water was produced by the melting of the top layers of the permafrost during summer and that the underlying chalk was rendered impermeable by being frozen. A more plausible explanation is that as the chalk scarp receded due to weathering and erosion, the water table gradually dropped. Thus a headstream on the chalk dip slope that was originally fed by a spring when the water table was high, has subsequently been left high and dry as the water table has been lowered. Today we often see intermittent streams in chalk country, known as bournes or combes, where the flow is controlled by the seasonal rise and fall of the water table.

Where the chalk bedding planes are inclined, the topography is characterized by steep scarps and gentle dip slopes. The water table will rise under the scarp crest and then dip steeply down to the base of the scarp, emerging at the junction between the Chalk (and Upper Greensand) and the Gault Clay. Water percolating through the chalk will issue at the junction with the impermeable clay as a scarp foot spring. At the foot of the chalk dip slope, springs will appear where the water table emerges at the junction with the overlying Palaeocene beds.

The surface of the chalk is often pitted with small solution hollows that may have developed along fissures or joints that have been widened by percolating rainwater. They are sometimes covered with a permeable cover of sand and gravel deposited during the Palaeogene or Pleistocene. Many old chalk pits display sections through these infilled pipes or fossil swallow holes. On the east side of Freshwater Bay, the chalk cliffs are capped with brickearth that has subsided into solution pipes, some of which are exposed as the cliffs have been cut back by weathering and marine erosion.

The grassland of a classic chalk down, here on the western end of Tennyson Down.

CHAPTER 8

How the Coastline is Shaped

Sea level has been relatively stable since the Flandrian marine transgression between 12,000 and 6,000 years ago, when the present outline of the Isle of Wight began to take shape. One of the most obvious manifestations of the sea level rise during the early Flandrian was the drowning of the river estuaries along the north coast of the island. The Newtown estuary is a classic drowned river mouth (called a ria by geomorphologists), with its tidal tributary channels extending several kilometres inland. Most of these estuaries have now been partly infilled with sediment brought down by rivers, creating alluvial floodplains and extensive intertidal salt marshes and mudflats. Below the present river mouths there are also buried channels dating from Devensian times, when sea level was much lower (*see* Chapter 6).

When we look at the geological map of the Isle of Wight, it is obvious that there is a direct correlation between relatively hard rocks like chalk and limestone and headlands such as the Needles, Culver Cliff and the Foreland. In contrast, large embayments, for example Brighstone Bay and Sandown Bay, are cut into softer clays and sands. These features result from marine erosion, where the sea actively wears down the coastal rocks in various ways.

Hydraulic action involves the waves crashing against the rocks and exerting pressure on joints and bedding planes so that they open up and eventually loosen blocks of rock. Waves armed with rock fragments also act like sandpaper to wear down the rocks in a process known as marine abrasion. Furthermore, when the sea is in contact with limestone or chalk then chemical weathering takes place, as calcium carbonate is removed in solution. The Bembridge Limestone is pitted by marine solution on the foreshore at Black Rock.

Much of the material eroded from the cliffs is later transported by longshore drift to be deposited on beaches and across river mouths, where spits may build up. The Duver is a sand spit extending south across the Eastern Yar, behind which is salt marsh and a tidal creek. Sand spits have developed on either side of the Newtown river estuary, protecting the salt marsh from the sea.

Blocks of Ferruginous Sandstone litter the beach in Compton Bay, where chalk cliffs extend into the distance.

The beautiful chalk stacks that form the Needles, with the steep northerly dip visible far left and far right.

Cliffs and Stacks

Steep cliffs are found where resistant rocks outcrop and they are usually accompanied by a wave-cut platform, which develops as the cliff line recedes. Whitecliff Point has a well-developed shore platform below its chalk cliffs, and on either side of Freshwater Bay the foreshore has been eroded into a rock platform at the base of the receding cliffs. Evidence for cliff recession is also provided by the sea stacks and arches that extend seawards from headlands. On the east side of Freshwater Bay the Mermaid Rock and Stag Rock are prominent chalk stacks and, between them, the Arch Rock stood until 1992, when it collapsed in a storm. Both the stacks and the arch have been shaped from hard, resistant chalk reinforced by layers of flint dipping at 65–70 degrees north. The Needles provide a classic example of multiple stacks marking the line of the chalk headland that during the Devensian would have continued westwards to Old Harry stacks off Ballard Down on the Isle of Purbeck.

Where softer clays and sands outcrop on the coast, the cliffs tend to be less steep and they are often subject to slumping as for example, at Bouldnor cliffs (Bembridge Marls) and Brook Undercliff (Wessex Formation). In Alum Bay and Whitecliff Bay, the Palaeocene strata are almost vertical, and marine erosion, weathering and slumping have sculpted the cliff face into intricate patterns. The slightly harder sandstones protrude as upright ribs whilst the softer clays have been excavated, forming deep gullies.

Cliff Movements

The work of the sea undermining cliffs is often accompanied by coastal landslipping, particularly around the south coast of the island. These mass movements can be classified as rock slides, mudslides, rock falls or rotational landslips.

Rock falls developing within the chalk cliffs along the edge of Compton Down as fissures open up in the steeply dipping strata.

Rock slides usually occur where the bedding is dipping seawards so that a competent, well-jointed rock layer will tend to move downslope when an underlying clay is lubricated by copious groundwater.

Mudslides tend to take place in soft, unconsolidated sands and clays after heavy rainfall, when they move downslope in lobate form. Blackgang Chine and the Bouldnor and Hamstead cliffs have well-developed mudslides that extend into the sea.

Rock falls are mostly confined to harder, well-bedded strata. The chalk cliffs on the edge of Compton Down have developed rock falls as fissures open up in the steeply dipping strata. In 1928, an estimated 100,000 tonnes of Chalk and Upper Greensand on Gore Cliff collapsed onto the old coast road, which now terminates

RIGHT: **Multiple mudslides pour over the sandstone ledges in the cliffs north of Blackgang.**

BELOW: **A mudslide of London Clay runs out into Whitecliff Bay, during the 1990s when it was more active than it is today.**

in a car park from which the rock debris can be viewed. About 500m to the northwest is the site of a major landslide that occurred in 1978, which destroyed Cliff Cottage and South View House. The saturated Gault Clay is transformed into a slippery slope over which rock debris slides down to the shore, where it is slowly removed by the waves.

Rotational landslipping can best be seen at the Undercliff between Ventnor and St Lawrence. The Lower Chalk and Upper Greensand form a steep cliff some 200m above sea level, overlooking a wide area of broken and slumped arcuate slices of the cap rock that have moved downslope over impermeable Gault Clay lubricated by groundwater seepage. It is possible that this rotational landslip may have developed under periglacial conditions during the Devensian since there have been numerous slides over a long period of time; although in recent decades the undercliff has remained relatively stable, the process of cliff recession is constantly at work.

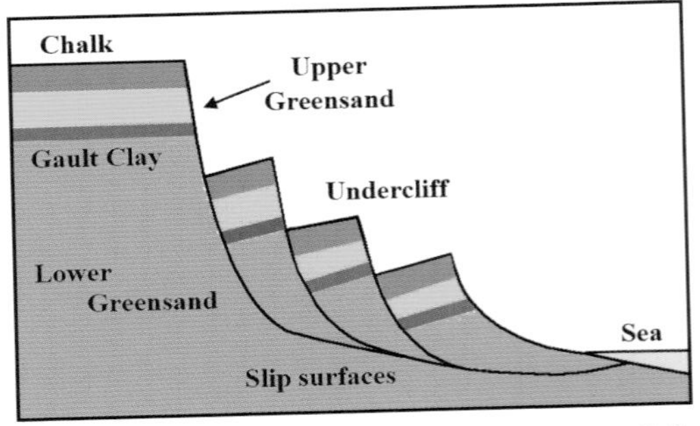

ABOVE: **Rotational landslips at the Ventnor Undercliff.**

RIGHT: **Chalk landslip at the southend of Gore Cliff, where the Upper Greensand forms the cliff face.**

BELOW: **View of the Undercliff from the chalk plateau near St Lawrence looking towards Binnel Point, showing the slumped masses of Upper Greensand and Gault Clay now covered with scrub vegetation.**

Movement in the Undercliff

Parts of Ventnor extend over the complex of landslides forming much of the Undercliff along the island's southern coast. Where large landslide blocks move slowly outwards, it is common for miniature rift valleys to develop behind them, when wedge-shaped slices (known as grabens) sink gently into the widening gaps. The step in the ground seen in this view developed progressively during the 1990s. The higher block, on the right, had moved almost horizontally, towards the right and out towards the coast. The lower block, on the left, had therefore moved downwards because it was on a subsiding graben, bounded also by another slip surface off the picture to the left. These landslides are still active, but extensive engineering works have greatly reduced the scale of current ground movements.

The Seabed Beyond the Island

We know that towards the end of Devensian times (late Pleistocene), the Isle of Wight was connected to the mainland as sea level dropped some 45m below its present level. The ancient Solent river with its tributaries, including the Medina, Frome, Stour and Beaulieu, drained the lowland in what is now the Solent-Spithead channel. But what rocks lay beneath this early river system that has long since been submerged beneath the waves? Geological outcrops do not, of course, end at the coastline: from the shoreline you can often see reefs of resistant hard rocks extending seawards below the surface of the sea.

The sands and clays of the Solent Group that occupy much of the northern half of the island extend across to the Hampshire mainland but marine erosion has removed them along the whole of the Solent-Spithead channel. Here the underlying Barton Beds outcrop along the submerged valley of the Solent river and are only seen onshore in Alum Bay and in the type locality at Barton-on-Sea near Christchurch.

continued on p.85

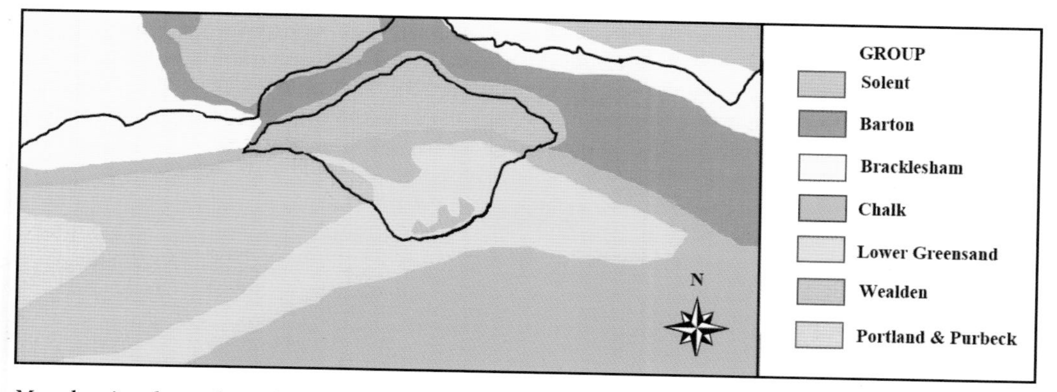

Map showing the geology of the seabed around the Isle of Wight, based on BGS Special Sheet: Isle of Wight. Simplified Bedrock Geology.

Itinerary 17: Windy Corner (below Gore Cliff) to St Lawrence via the Undercliff

Walking distance: 4km

From Niton village, drive along Sandrock Road to Windy Corner car park [SZ 494758], where the 1928 landslide destroyed the old coastal road.

From here there is a fine view of Gore Cliff, formed of horizontally bedded Upper Greensand reinforced with chert bands and capped by the Lower Chalk. At the base of the cliff, the impermeable Gault Clay creates a seepage zone where groundwater reaches the surface, causing both cliff recession and landslipping. Take the steep footpath down to Watershoot Bay, where the beach is a mass of boulders brought down by landslipping. The boulders are mainly from the Upper Greensand and the Glauconitic Marl at the base of the Lower Chalk. Greensand blocks can be recognized by their chert concretions set in grey siltstones, whilst the Glauconitic Marls are composed of buff sandy clay with dark-brown phosphatic nodules.

Take the footpath leading eastwards to Knowles Farm and St Catherine's Lighthouse, which is situated on the low coastal terrace where it is backed by the hummocky landslipped slope beneath the cliff edge of the chalk downs. The farm buildings are surrounded by stone walls enclosing large fields that provide good-quality grazing land.

Join the coast path where the receding cliffs are cut into chalk rubble and boulders of Upper Greensand. Among the rock-debris-strewn shore patches of Gault Clay are exposed. Continue past Castle Haven to Reeth Bay, where blocks of red-brown ferruginous sandstone lie on the beach where they have fallen from the overlying Carstone. West of the slipway, the toe of the debris apron of a coastal landslip can be seen.

Next, walk on to Puckaster Cove, where the Sandrock Formation is exposed in the low cliffs. Here the sands exhibit a variety of sedimentary structures, including cross-stratification, graded bedding and flaser bedding. The last of these is formed of alternating sand and mud layers typically deposited in an estuarine or tidal environment.

Continue to the eastern side of Binnel Bay, where debris from the Undercliff landslide complex reaches down to the shore. The slumped cliffs are a chaotic mass of boulders produced by mass movement, including debris flows and mudflows. The boulders are from the Upper Greensand and Glauconitic Marl, and the slumping low buffs behind the beach are formed of Gault Clay. Some of the Upper Greensand boulders contain weathered-out bivalves (*Exogyra* and *Neithea*) and ammonites such as *Mortoniceras*.

From Binnel Point [SZ 523758] there is an escape route via a pathway up through the scrub vegetation of the Undercliff to the A3055 on the western side of St Lawrence. Alternatively, you can continue eastwards to Woody Bay, where back-tilted slipped masses of Upper Greensand form the cliffs, and then take the path up to St Lawrence village, nestling on the Undercliff. You will probably need to arrange some transport from here back to the car park at Windy Corner!

ABOVE: **Rotational landslip below Gore Cliff, where the chalk has slipped over underlying Gault clay. The old coast road is blocked and used as a car park.**

BELOW: **St Catherine's Lighthouse is built on a coastal terrace backed by the landslipped hummocky slope beneath the edge of the chalk plateau.**

Itinerary 18: Alum Bay to Totland Bay over Headon Hill

Walking distance: 4km

Car parking is available next to the Needles Pleasure Park [SZ 306854], and from there you can descend through Alum Bay Chine to the beach. Walk northwards towards Hatherwood Point, from where it is possible to ascend the side of Headon Hill. The structure of the hill is clearly visible from this seaward side as there are prominent beds of horizontal limestone separated by vegetated sandy clay slopes.

At beach level, the Becton Sand (Upper Barton Group) is exposed, but this is overlain by marls that form the base of the Headon Hill Formation. These marls contain various gastropods, including *Viviparus* and *Galba*. The How Ledge Limestone forms the first terrace and above this is another marl bed, containing a wide variety of both bivalves and gastropods but, unfortunately, much of this unit is obscured by mudflows. The prominent cliff of Hatherwood Limestone marks the edge of the second terrace. Irregular beds of lignite occur near the base of the limestone and they have yielded mammalian vertebrae and fragments of turtle. A further slope rises to the third terrace, where the Lacey's Farm Limestone forms the highest cliff face. The topmost beds of the Headon Hill Formation are poorly exposed on top of the hill, which is capped by Plateau Gravels (Lower Pleistocene fluvial deposits), which are mostly covered with bracken and heather, although there are some partly overgrown abandoned gravel pits.

The heather-covered slopes on the top of Headon Hill, which is covered with Plateau Gravels left by late Pleistocene rivers. The view northwards is across Totland Bay and Colwell Bay to Cliff End.

Walk over to the north side of the hill, where the Bembridge Limestone rests on the Headon Hill Formation and forms a line of cliffs overlooking Totland Bay. The limestone contains freshwater gastropods that can be examined in an old quarry at SZ317863. Follow the coast path down to the seafront at Totland, taking in the view northwards across Totland Bay and Colwell Bay towards Fort Albert at Cliff End. Hurst Castle on the mainland can be seen on the shingle spit across the narrow strait in the approaches to the western Solent.

Headon Hill seen from the sea. Cliffs of Becton Sand can be seen at the base and Hatherwood Limestone midway up the hill. The Lacey's Farm Limestone forms a prominent cliff feature near the summit.

Itinerary 19: Nettlestone Point to Bembridge Harbour and the Foreland

Walking distance: 8km

Nettlestone Point is on the northeast coast of the island, backed by Seaview village, where parking is available.

The Nettlestone Grit outcrops as ledges along the shoreline; this is a resistant indurated ferruginous sandstone that occurs near the top of the Headon Hill Formation. At low tide you can walk across the sandy beach of Seagrove Bay to Horestone Point, on the northern side of which a rotational landslip has caused the Bembridge Limestone and overlying marls to subside on the shore. A fault brings the Nettlestone Grit up to form the promontory, which slopes down to a much dissected and eroded shore platform.

Continue southwards across Priory Sands, overlooked by the low, wooded cliffs of slumped clays that dip beneath the Bembridge Limestone and Bembridge Marls. Inland, the latter are covered by Pleistocene marine gravels (Wootton Gravel Complex), which cover the St Helen's plateau.

Nodes Point lies at the southern end of Priory Bay, where the Bembridge Limestone extends seawards to form an extensive shore platform backed by slumped cliffs of Bembridge Marl. South of Node's Point, the 6m-thick Bembridge limestone is interbedded with clay bands and it is much dissected along joints and bedding planes. These ledges dip gently south towards the tower of St Helen's Church and the axis of the Bembridge syncline, which runs approximately east–west through Bembridge Harbour. The far side of the syncline is seen at Tyne Ledge off Bembridge Point, where the limestone beds dip northwards.

The coastal area around Bembridge.

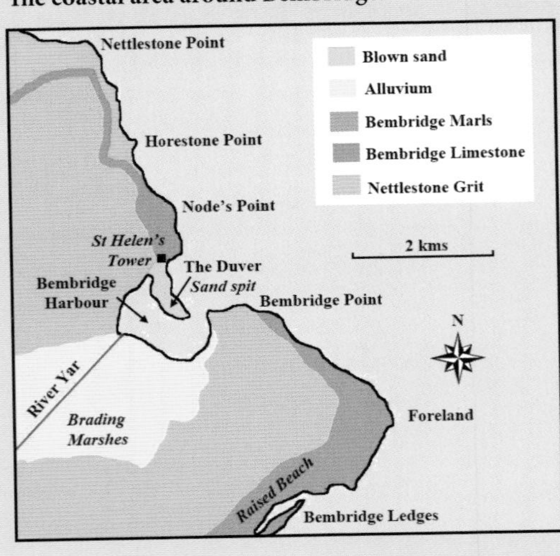

The Duver

St Helen's Church tower, on the sand spit known as the Duver, is in the care of the National Trust. The structure survives from the thirteenth century, when the church served a Benedictine community. Later it fell into disrepair, and in 1719 the tower was painted white to act as a seamark for Royal Navy ships that anchored in St Helen's Roads to take on provisions.

From the tower, walk along the seaward edge of the spit, past groynes protecting the sand and shingle beach. Behind the beach the dunes are stabilized by marram grass; the sandy scrubland was once the site of the Royal Isle of Wight golf course, established in late Victorian times for men only. Proximity to Osborne House encouraged royal visitors, but its fortunes declined in post-war years and in 1961 the land was given to the National Trust as an open space for recreation.

Walk over the Duver and follow the stone causeway, or old mill dam, that crosses the entrance to the tidal inlet known as the Old Mill Ponds – a reference to the former tidal mill that stood at the southwestern end of the causeway. Spartina grass and salt marsh vegetation cover the mudflats within the lagoon.

Bembridge Harbour and the Foreland

The coastal footpath continues along the bank around the south side of Bembridge Harbour to Bembridge Point, where a spit protrudes northwards opposite the Duver spit, confining the mouth of the River Yar. The path follows the shingle beach to Tyne Ledge and to the lifeboat station pier at Ethel Point. Bembridge Limestone forms reefs along the shoreline that can be traced from Tyne Ledge to Bembridge Ledges on the Foreland, the most easterly point on the island. The limestone forms a series of gently inclined structural platforms separated by thin interbedded clays that have been much eroded along the strike of the beds.

From the lifeboat station, a path follows the shoreline around the Foreland to Foreland Fields [SZ 652872], opposite Black Rock Ledge, where low cliffs are formed of Pleistocene gravels on a bench cut into Bembridge Marls. This is the Bembridge Raised Beach, which extends back to the base of a degraded cliff about 15m above the present sea level. A loamy brickearth is banked up against the palaeocliff, which forms the southern boundary of the marine gravels that were deposited during the Ipswichian interglacial, when sea levels were higher than today.

continued from p.81

The backbone of the Isle of Wight formed by the east–west monoclinal fold of the chalk ridge can be traced from the Needles to Ballard Down on the Isle of Purbeck. This barrier was breached as sea level rose with the Flandrian transgression some 12,000 years ago. From Culver Cliff, a similar narrow outcrop of chalk runs southeastwards into the English Channel, where it broadens out into an extensive area to the south of the Isle of Wight.

The Lower Greensand that forms much of the southern part of the island extends seawards off the south coast, where its relatively soft sediments have been eroded to form the St Catherine's trench over 80m deep. The underlying Wealden Beds outcrop along the southwest coast and continue in a wide belt towards the Dorset coast, where the underlying Portland and Purbeck strata are present. It is this Jurassic succession that forms the concealed basement beds on which rest the Cretaceous and Palaeogene sediments of the Isle of Wight.

Much of the seabed around the island is covered with coarse sand and gravel derived from the palaeo-Solent river system. These sediments clearly mark the course of the river during the Pleistocene as it flowed west to east along the Solent-Spithead axis and then turned south to join the Channel River. Coarse sediments are also found in channels running north to south off the Hampshire coast, and these are thought to belong to tributary rivers that cut through the Purbeck-Wight monocline before it was finally breached by the rising sea level.

Changing Sea Levels: Eustatic and Isostatic

During the Pleistocene period, sea level fluctuated between glacial lows and interglacial highs. In the Devensian glacial phase, sea level was much lower than it is today because vast quantities of sea water were locked up in glaciers and ice sheets. Since then, sea levels have gradually risen because land-based ice is melting and ocean waters are warming and consequently expanding.

It is interesting to note that when sea-based ice melts there is no increase in the volume of sea water. An iceberg floats because it is less dense than water and because it displaces an equivalent mass of water, so there is no rise in sea level as it melts. However, when land based Arctic ice melts, the albedo effect comes into play. This is the ability to reflect the sun's rays back into the atmosphere. Since ice and snow are very good at reflecting sunshine, when they melt and expose the land surface it then warms up.

Global change in the volume of sea water, due to post-glacial melting, is referred to as a eustatic change in sea level. By contrast, during glacial phases when the land is covered by great thicknesses of ice, the weight of ice depresses the surface of the Earth's crust and later, when the ice melts, the crust rebounds. This causes an isostatic sea level change since the land is rising, not the volume of water increasing. Orogenic uplift can also decrease sea level by raising the land and changing the size of the oceans.

Low sea level during the Devensian glaciation enabled rivers to cut their channels down to a low

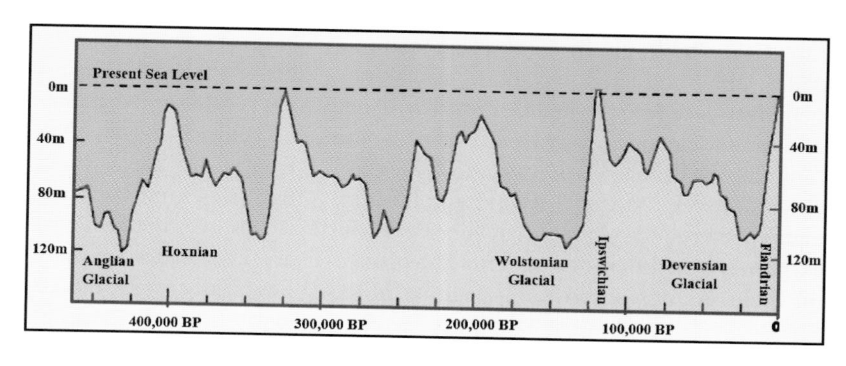

Fluctuations in sea level during the last half a million years.

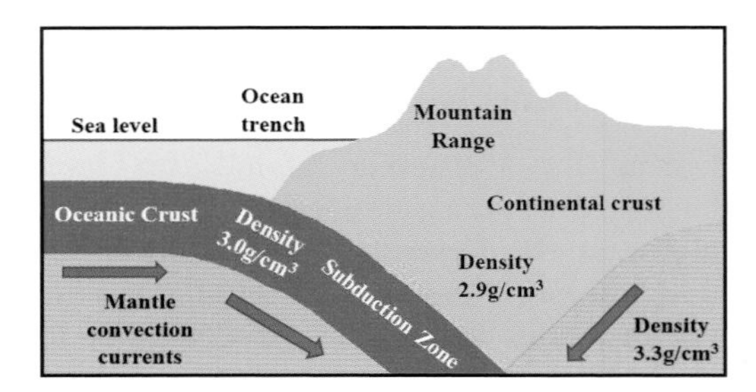

Isostatic equilibrium demonstrated by the lower-density continental crust within a mountain range 'floating' on denser mantle material.

base level, but as eustatic sea level increased the lower tributary valleys and river estuaries were flooded to produce rias, as seen on the north coast of the Isle of Wight. In glaciated regions, the drowned lower parts of glacial valleys form fjords, best known in Norway.

Isostatic adjustment after the Devensian glaciation has produced raised shorelines up to 14m above present sea level in the Grampian highlands, at the centre of uplift. Further west, on the Isle of Mull, raised beaches are around 10m above sea level. These distinctive landforms are remnants of former wave-cut platforms backed by degraded cliffs, having been abandoned as the land has been uplifted.

Isostacy is the term given to the principle of buoyancy; when an object is immersed in a fluid, it is buoyed with a force equal to the weight of the displaced fluid. In geology isostacy can be observed where the Earth's crust 'floats' on the underlying denser mantle in a state of isostatic equilibrium.

Regions that are tectonically stable tend to be in isostatic equilibrium, but where mountain ranges are built up by converging tectonic plates, the subducted mountain roots may penetrate deep into the mantle and be held out of isostatic balance. As erosion proceeds, the mountains will be lowered, become lighter and tend to rise to restore isostatic equilibrium; equally, surrounding areas will subside as the extra weight of the eroded material is deposited on them. As noted above, the loading of ice sheets can also produce an isostatic response.

Time and Tide on the Foreshore

The shoreline around the Isle of Wight can be a dangerous place for unwary geologists seeking out rock formations far from the maddening crowd! Many readers will be familiar with the workings of the tides, but from a safety point of view, and to avoid being cut off by the tide, it is well worth thinking about the tidal regime.

Firstly, before starting an excursion, you must be aware of the state of the tide. Read the tide timetable, which will show the time of high and low water and the height of the tide. The tide times are usually given in Greenwich Mean Time so make sure that you adjust for British Summer Time if necessary; miscalculation of one hour difference can jeopardise your schedule and your safety. Spring tide occurs twice every lunar month (twenty-eight days), when the moon, Earth and sun are in alignment at new and full moon. This tide produces the highest level of high tide and the lowest level of low tide; in other words, the tidal range is greatest at springs. By contrast, the tidal range is smaller when neap tide occurs seven days after springs, when the sun and moon are at right angles to each other at half moon.

Many of the beaches on the island are composed of sand and shingle derived from local rocks. The coarser pebble material is moved up the beach by the swash while the finer sand and mud is moved seawards by the backwash. During winter storms, a shingle ridge may build up at the head of the beach. Longshore drift operates on many beaches where the wave fronts are aligned at about 45 degrees to the shoreline, hence the importance of groynes on holiday beaches to prevent sand being washed further along the shore.

While on the beach, take the opportunity to study patterns in the sand. Symmetrical ripples are seen where shallow tidal water oscillates sand back and forth, creating sand ridges up to 5cm high. Asymmetrical ripples are produced when there is a strong directional flow. The current moves sand particles up the gentle stoss slope on the up current side and

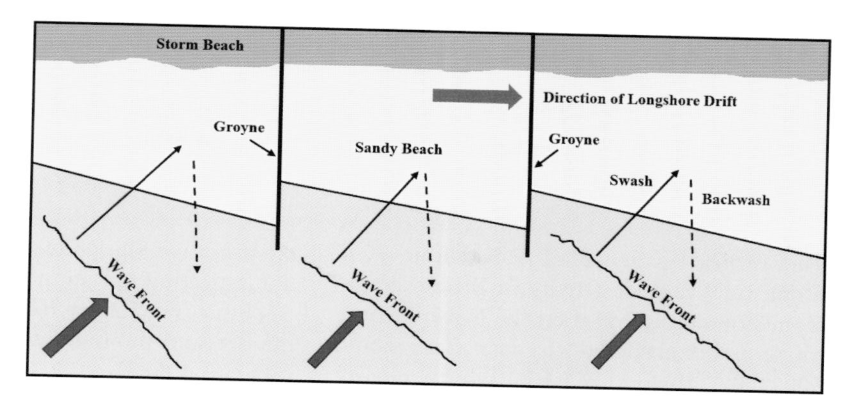

Beach development under the influence of longshore drift.

over to the steep lee slope, in a similar way to the formation of cross-stratified desert sand dunes. Next, at low tide, you may see water issuing from the shingle above the beach. This water will often cross the beach as a meandering rivulet or it may become braided and split into several smaller branches. Look carefully and you will see in miniature many of the features displayed by a mature river. There are undercut river cliffs. point bars, meanders and a delta where the stream enters a beach pool. These model features help in understanding the structures we see in sedimentary strata exposed in the field.

When moving from the beach to a wave-cut platform beneath the cliffs, we enter a more hazardous environment. This is hard hat territory below crumbling cliffs, and strong footwear is recommended for clambering on slippery rock surfaces that are washed by the tide. Nevertheless, the boulders and loose rocks on the foreshore that have fallen from the cliffs above can provide some useful study material. They may contain fossils or sedimentary structures that can be examined more easily at ground level than higher up the cliffs. Fallen septarian nodules can often be split open to yield crystalline minerals or well-preserved fossils. Where gently inclined rock ledges extend seawards beyond the cliffs, it, may be possible to see fossil shell beds on the rock surface that was once the seabed. The ledges form reefs that tend to slowly break up, littering the foreshore with rectilinear boulders.

A storm beach of cobbles and small boulders, nearly all of chalk or flint, in Alum Bay; the shingle grades seawards into pebble and sand, which can just be seen in the centre of the photograph between the foreground shingle and the distant chalk cliff.

CHAPTER 9

From Prehistoric Burials to Tudor Forts

The earliest evidence of Palaeolithic (Old Stone Age) human activity in Britain comes from stone tools found on the Norfolk coast. These are considered to date from about 900,000BP (before present, where present is defined as 1950), when the Weald-Artois anticline joined southeast England to the continent and the Isle of Wight was part of the mainland. However, the extreme cold of the Anglian glaciation would probably have caused emigration to the continent, and it was not until the warmer Hoxnian interglacial that a Palaeolithic population was re-established. Even so, the succeeding glacial phases of the Wolstonian and Devensian must have made life difficult in the periglacial areas south of the ice sheets.

The Earliest Inhabitants

By the time the ice sheets receded, nomadic hunter-gatherers were inhabiting southern Britain. Evidence on the Isle of Wight is provided by numerous discoveries of flint tools in the Plateau gravels, river terrace gravels and brickearth, and some of these are displayed at Carisbrooke Castle. Around 6000BC, during the Mesolithic Age, a settlement existed near Bouldnor. This was discovered in 1999 by divers in the shallow waters of the Solent off Bouldnor Cliff, where wooden timbers, cooking pits and flint tools were preserved within the sea floor muds. There appears to have been a thriving community here, probably living on fish and game close to what is now a submerged forest. The settlement was eventually overwhelmed by rising sea levels during the Flandrian marine transgression.

By about 4300BC, the Neolithic (New Stone Age) had begun, with the development of primitive agriculture arising from the domestication of animals and plants. This Neolithic revolution led to a more settled lifestyle and the building of stone structures venerating the dead. It is also noticeable that most standing stones and burial mounds were sited on open chalk and limestone uplands well above the damp, densely forested lowlands.

On the Isle of Wight, there are only a few preserved Neolithic sites. The Longstone [SZ 407842] near Mottistone is a megalith standing almost 4m high, formed of Upper Greensand. There is a second smaller stone lying on the ground and they were probably part of the revetted stone entrance to an earthen long barrow (burial mound). The structure was aligned east–west, possibly for use by people worshipping the sun and moon. The Longstone is now a National Trust monument and you can visit it by climbing the gentle footpath from Strawberry Lane out of Mottistone village.

The Longstone, a Neolithic megalith, or standing stone, near Mottistone.

There is a Neolithic long barrow [SZ 351857] on Afton Down near Freshwater Bay golf course that is located in the middle of a Bronze Age round barrow cemetery. The Afton long barrow is orientated east–west and has a ditch on either side of the long axis but there is no evidence of a stone-built entrance. About 1.5km to the west, on Tennyson Down, there is a Neolithic mortuary enclosure [SZ 336855] marked as an 'earthwork' on the 1:25,000 Ordnance Survey map. The enclosure is surrounded by an oblong-shaped embankment and ditch. This type of burial site is quite rare nationally and is a scheduled ancient monument protected by Historic England.

The Bronze Age in Britain lasted from around 2200BC to 750BC, and this was a time when megalithic structures such as Stonehenge were erected. The Beaker people brought metalworking techniques, including the making of bronze by mixing copper with small amounts of tin. Bronze Age copper mines existed at Parys Mountain on Anglesey and on the Great Orme, Llandudno, and tin ores were worked in Devon and Cornwall.

At least 200 Bronze Age burial mounds have been identified on the Isle of Wight, most of which are on the chalk downland. These round barrows are marked as tumuli on the Ordnance Survey map. Some burial mounds are isolated, such as the one on Headon Warren [SZ 312858], while others are found in groups, as on Ashley Down [SZ 577877], Brook Down [SZ 390851] and Mottistone Down [SZ 406847]. Some sites are merely ring ditches with a central crater, which have suffered the predations of medieval barrow robbers and nineteenth-century excavators or have been ploughed over for centuries. It is of interest to note that although the dead were generally buried in round barrows, during the latter part of the Bronze Age bodies were cremated and the ashes stored in funerary urns.

Around 750BC, iron-working techniques were introduced to Britain from the Mediterranean region and since iron was stronger and more ubiquitous, it soon replaced bronze in the production of swords, spears and agricultural tools. Hence this technological revolution marks the beginning of the period known as the Iron Age in Britain. By the time of the Roman invasion in AD43, there were around 2,000 hilltop fortifications in Britain although only one survives on the Isle of Wight. This is the Chillerton Down promontory hill fort, standing above the chalk scarp. Today, a prominent tall television mast stands on the chalk plateau at 167m above sea level overlooking the ancient defended site. Although there is little evidence on the ground of Iron Age settlement, the lowland forests would have been slowly cleared for arable and pastoral farming. Iron axes were used to clear the deciduous woodland and iron ploughs made for deeper more efficient cultivation of the soil.

Roman Invaders

Some 200 years before the Roman invasion of AD43 the coastal areas of southern Britain saw an influx of people from Gaul known as the Belgae, who enhanced agricultural development and brought advanced pottery-manufacturing skills. There are Roman villa sites on the Isle of Wight (Roman Vectis) where late Iron Age pottery has been found, which suggests that they were occupied continuously from pre-Roman times.

The Roman villa at Brading [SZ 598862] is open to the public and has some well-preserved mosaic floors with geometric patterns and depictions of Roman gods and goddesses. Extensive use was made of local building stone, including chalk, flint, greensand, and Bembridge Limestone for roof slabs. The remains of field systems with low banks or lynchets associated with the villa can be seen on Brading Down. The villa was situated on the Ferruginous Sandstone, a well-drained south-facing site, sheltered to the north by the chalk ridge of Brading Down. Another Roman villa, in the southern residential area of Newport, was discovered in 1926 when the owners of a house in Cypress Road started to dig the footings for a garage. A beautifully preserved hypocaust system for underfloor heating, a bath suite and mosaic floors were uncovered and are now on display to the public.

In total, there are at least seven villa sites on the island and all were owned by wealthy Romans closely involved with the agricultural economy. Outbuildings included slave quarters, granaries and livestock enclosures. There is also evidence of pottery kilns for making Vectis ware and flagons for wine storage. The light, chalky soils and south-facing scarp slope of the central downlands would have been ideal for viticulture.

Anglo-Saxons and Normans

The Roman legions withdrew from Britain around AD410 as the Roman empire began to decline. At the same time, Germanic tribes began to invade eastern Britain and gradually these became the Anglo-Saxon settlers. They were essentially farming communities establishing settlements where fertile land and water supplies were available. On the Isle of Wight, scarp foot spring lines provided water, while there was access to the chalk downs for sheep grazing and to the heavier soils for arable cultivation on the Gault and Lower Greensand. Villages such as Arreton, Chillerton and Shorwell occupy typical Anglo-Saxon scarp foot sites sheltered beneath the chalk downs.

Although Christianity was first introduced into Britain by the Romans, it was not until the seventh century that many pagan Anglo-Saxons were converted, and churches and monasteries were gradually established. St George's Church at Arreton has a Saxon wall, doorway and window arch, and in All Saints Church at Freshwater there are Saxon quoins (corner stones) on three of the nave piers. Other churches on the island are dedicated to Saxon saints and martyrs and rebuilt in Norman times on the site of a Saxon church; they include St. Boniface (Bonchurch), St. Mildred (Whippingham) and St Edmund (Wooton).

The Norman conquest of 1066 led to the subjugation of the native population as William the Conqueror appointed his barons to positions of authority. They occupied landed estates and manor houses, and a series of castles was built to defend and control the country. A survey was ordered to find out ownership of land holdings and their value and the information was recorded in the Doomsday Book. Details of each parish were entered in the book; entries for the Isle of Wight parishes were included in the Hampshire records.

There is a splendid example of a Norman Castle at Carisbrooke, complete with motte

The gatehouse at Carisbrooke Castle, a Norman fortification that stands on a chalk spur overlooking Newport and the Medina valley.

View from the battlements of Carisbrooke Castle over Newport and the Medina valley.

and bailey, curtain walls and gatehouse, that dates from the early twelfth century. The castle stands on a late Saxon fortified earthwork on a chalk spur overlooking the Lukely Brook, a tributary of the Medina. Local chalk and flint were used in the construction of the castle's walls and fortifications. The 'new port' for Carisbrooke was established near the tidal limit of the River Medina at Newport around 1190. The church of St John the Baptist at Yaverland was originally built around 1150 as a private chapel. The nave and chancel, with its Perpendicular-style windows, reflect the Norman ecclesiastical architecture.

Middle Ages and Beyond

The island continued to be governed by a Norman lordship throughout the reigns of the Plantagenet kings (1154–1485) but defence of the realm against French raids became an important issue. By the time of Henry VIII in the early sixteenth century, the construction of several coastal forts was undertaken. Yarmouth Castle is a Tudor artillery fort built in 1547 to protect Yarmouth harbour and the Solent approach to the mainland. Blocks of Bembridge Limestone from Quarr Abbey (which had been dissolved by Henry) were used in the building of the castle. The fortifications were redeveloped and improved over the succeeding centuries and it was last used by the military during World War II.

Newtown, on the northwest coast, has an interesting Norman history. It was designed by the Bishop of Winchester and granted borough status in 1256, but it suffered considerable damage from French raids in the fourteenth century and declined rapidly thereafter, although a town hall was built in the late 1600s. Nevertheless, Newtown survives as a testament to medieval town planning, with its rectilinear street pattern, stone quay and adjacent salterns, which are the last remnants of a once-thriving medieval salt industry.

The seas around the south coast of the Isle of Wight can be dangerous, particularly for sailing ships in the medieval period. Hence during the early fourteenth century, a lighthouse with an adjoining oratory (chapel) was built near to the summit of St Catherine's Hill [SZ 493771]. This was done as an act of penance by Walter de Godeton for plundering wine from a shipwreck in Chale Bay. You can visit the lighthouse, known as the 'Pepperpot', by climbing the path from the car

Part of the wall at Carisbrooke Castle, showing the use of rough-hewn chalk along with some dark flint nodules that display conchoidal fracture.

park alongside the A3055 above Blackgang. Note the strip lynchets on the hillside used for cultivation in medieval times, and the Bronze Age barrow near to the lighthouse. The oratory has completely disappeared now. The lighthouse was not very successful, partly due to its location, which was often shrouded in mist. The modern lighthouse is built below the undercliff on the coastal terrace.

The oratory of St Catherine's built on the summit of St Catherine's Hill.

Prehistoric Monuments in Britain

Camps Some of the oldest prehistoric remains in Britain are the Causewayed camps, dating from around 3500BC in the Neolithic period. They consist of several concentric rings of banks and ditches enclosing a central area that was probably a multi-purpose gathering place. The ditches were bridged by earthen ramps or causeways, hence the name, although the term 'camp' is a misnomer since the enclosure was most likely used as a ceremonial or feasting area or even as a livestock pen. Windmill Hill near Avebury in Wiltshire is the best preserved example of this type of monument.

Long barrows These communal tombs holding the remains of up to fifty people are considered to be contemporary with Causewayed camps. The barrows were up to 120m long and orientated east–west, usually tapering towards the west. They were the site of religious activities and their orientation was probably associated with the cult of the rising sun. The West Kennet long barrow, two kilometres south of Avebury, is one of the largest and most impressive stone chambered tombs in Britain.

Standing stones or megaliths These are often in the form of a stone circle, which have been constructed since Neolithic times. There are at least 900 stone circles still surviving in Britain, from Cornwall to the Orkneys. Salisbury Plain, with its undulating chalk landscape, was a focus of prehistoric people, and here there is a concentration of burial mounds, earthworks and circle monuments, the most famous of which are Stonehenge and Avebury.

Local sarsen stones were extensively used in building the stone circles. The sarsens are hard sandstone blocks that lie on the surface of the Marlborough Downs. When the chalk was covered by Palaeocene sands and gravels, these were cemented by the passage of silica-rich waters to form hard sheets of sandstone.

The twelfth-century Norman church of St John the Baptist at Yaverland, built largely of local stone.

During the periglacial conditions at the end of Devensian times, the sheets were broken up and loose blocks or sarsens were scattered over the chalk downs. Some of the larger sarsens are 9m tall and weigh around 25 tonnes. The inner circle of Stonehenge is constructed of smaller 'bluestones' (spotted dolerite) that come from the Preseli hills in Pembrokeshire. The question of how they were transported over 220km has been a controversial subject of debate for over a century. The stone circles were probably associated with religious ceremonies and they also had astronomical significance concerned with summer and winter solstices.

Round barrows A form of burial mound, round barrows were introduced to Britain from the continent by the Beaker People around 2200BC. They were built for the individual burials of important people and sometimes contained elaborate grave goods, such as bronze daggers, pottery jars and necklaces. These Bronze Age barrows are marked as 'tumuli' on Ordnance Survey maps and they are numerous on the chalk downlands of southern England.

Hill forts Dating from the Iron Age (750BC –AD43), these hilltop enclosures are usually earthworks in the form of circular banks and ditches. However, these so-called forts were not primarily defensive structures and may have been used for ceremonial or religious purposes and social gatherings. Over 3,000 hill forts were built in Britain during the Iron Age but many were abandoned after the Roman invasion. One of the most spectacular and well preserved structures can be seen at Maiden Castle in Dorset.

CHAPTER 10

Railways and Seaside Towns

The Victorian era brought many changes to the Isle of Wight, not least the coming of the railways and the development of seaside holiday towns. Railway mania reached the island with the opening of the Newport–Cowes line in 1862. Shortly afterwards, the ferry terminal at Ryde was connected to Shanklin, and by 1866 the line was extended to Ventnor. The Newport–Sandown line opened in 1875 and the Brading–Bembridge branch line in 1882. Then, in 1889, the Freshwater, Yarmouth and Newport line began operations. The network was finally completed in 1900, when the branch line was finished between Merstone, on the Newport–Sandown line, and Ventnor West. At this time there was a total of almost 90km of rail network operating both passenger and freight services. There was even a project to build a tunnel from Ryde to extend the railway to the mainland, but this was abandoned due to the onset of the World War I.

The Railways Act of 1921 aimed to integrate the various island railway companies into the Southern Railway, which made significant investment in the infrastructure and rolling stock. In 1948, the UK railways were nationalized, and British Rail was created. Within a few years, most of the island's lines

A view looking northwards showing the Isle of Wight heritage steam railway, green fields on the Palaeogene sands and clays, and beyond across the Spithead towards Portsmouth.

Needles Old Battery. A Victorian fortification built in 1862 to protect against a French attack that never materialised. Hence it became one of 'Palmerston's Follies' since it was he who commissioned it. Now it is a marvellous tourist attraction administered by the National Trust.

were closed and, shortly after the publication of the Beeching report (1963), the only line left open was that between Ryde and Shanklin, which was electrified in 1967 using the third rail system, and since then has equipped with sets of trains that had previously seen service in the London Underground.

But all was not completely lost, because in 1971 the Isle of Wight steam railway was opened and now runs for 9km between Smallwood Junction on the Ryde–Shanklin line and Wootton, via Ashey and Haven Street stations. The steam railway runs along the former Ryde–Newport route, which may eventually be extended towards Newport. This heritage line is a significant tourist attraction during the summer season and is staffed largely by volunteers. Some of the lines abandoned in the 1950s and 60s have been converted into cycle tracks; notably National Cycle Route No. 23 from Cowes to Shanklin via Newport, and the disused rail track from Freshwater to Yarmouth is also a cycle path.

There are a variety of sea transport links to the mainland, which operate from ports along the north coast of the island. Ryde has benefitted from its early nineteenth-century pier, which is 680m long and reaches across a wide beach at low tide. It still serves the present-day Isle of Wight railway and the passenger catamaran service to Portsmouth Harbour. A hovercraft service also runs from Ryde to Southsea. About 4km west of Ryde is the Fishbourne terminal, where vehicle ferries operate a service to Portsmouth. East Cowes at the mouth of the Medina river has a vehicle ferry sailing to Southampton and, in the far west of the island, the Yarmouth–Lymington crossing provides a relatively short link to the mainland.

Palmerston's Follies

Travelling across Spithead from Portsmouth to Ryde, you cannot fail to notice (unless it's foggy) Spit Bank Fort or No Man's Land Fort rising above the waters. These are Palmerston forts that were built during the 1860s on the recommendations of a Royal Commission to defend the approaches to the Solent and Spithead from French attacks. They were championed by Lord Palmerston, the prime minister, but were later described as 'Palmerston's Follies' because they soon became obsolete as the French threat passed and they were considered a great waste of money.

Even so, some forts were still used in World War II as observation posts. There are three groups of forts: one group stands in Spithead, guarding Portsmouth; another group is built along the east coast of the isle of Wight; and a third group stands on the northwest coastline. Bembridge Fort [SZ 624861] is a good example of a Palmerston Fort; built on Bembridge Down, this hexagonal fort was designed as a command and control centre and as a final retreat if the island was invaded!

Ventnor and the East Coast

On the south coast, Ventnor was a tiny fishing village at the beginning of the nineteenth century, but by the 1830s it had grown into a thriving seaside town with hotels and boarding houses for visitors. The main reason for its rapid development was the discovery that Ventnor's subtropical microclimate was ideal for invalids suffering from consumption, a common cause of death at the time. Sheltering below the chalk downs on the Undercliff with its mild winters and warm, pleasant summers, Ventnor became a fashionable health resort. A sanatorium was established in 1869 for the treatment of tuberculosis. The railway reached Ventnor in 1866, enabling wealthy Londoners and others to visit in larger numbers. There was even an 'invalid express', bringing consumptive patients from the ferry terminal at Ryde.

During World War I, Ventnor became a centre for wounded soldiers, but the fashionable health business never really recovered. After the war, holidaymakers arrived both by train and on the steam packet direct from Southend to the Ventnor pier. The RAF radar station above Ventnor was targeted during World War II and the town was also bombed. Post-war austerity and the rise of package foreign holidays adversely affected trade in many British seaside towns including Ventnor, and unfortunately it has never been able to regain its Victorian popularity.

Ventnor developed as a seaside health resort in the1830s. Its sheltered south facing aspect on the undercliff ensures an almost sub-tropical micro climate which still appeals to visitors today. This picture was taken in winter, which accounts for the lack of holiday crowds.

There are several other seaside resorts on the east coast of the Isle of Wight, including Ryde, Sandown and Shanklin. They developed in Victorian times as ferry services and the railways made them accessible to holidaymakers from the mainland. All these towns benefit from extensive sandy beaches and safe bathing, which are essential for family holidays with children. Over the years, numerous hotels and guest houses have been established to provide tourist accommodation plus restaurants, cafés and entertainment facilities. Sandown has the Dinosaur Isle geological museum and the Zoological Gardens to attract visitors.

Cliffs of the Ferruginous Sands provide sand to the beaches between Shanklin and Sandown.

Sailing Capital

The largest port on the island is Cowes, on the west bank of the Medina estuary, which is linked by a chain ferry to the smaller town of East Cowes on the east bank. In 1815, the Royal Yacht Squadron was established and Cowes became a base for international yacht racing. Royal patronage, initially provided by George IV, promoted the sailing regatta in what eventually became the Cowes Week, involving hundreds of competitors from weekend sailors to international racing crews. As Cowes developed as a world-renowned sailing centre, its marine operations flourished; boatbuilding yards, sail-making works, engine repair workshops, chandlers, marinas and yacht design offices were established. Tourism is an important source of income for many businesses in Cowes and sailing enthusiasts arrive in large numbers for the annual regatta in early August.

The long drowned valley, known as a ria, of the River Medina, which provides the perfect sheltered harbour for the yachting fraternity at Cowes.

CHAPTER 11

Rural Land Use

Geology, topography and climate exert considerable influence on agricultural land use on the Isle of Wight. The climate is generally mild, with average temperatures ranging from 6°C in January to 18°C in July, with around 800mm of rain per year.

Agriculture

On the chalk downlands, the calcareous soils are well drained, relatively thin and low in nutrients, so they support short grasses and many species of wild flowering plants including orchids, bird's-foot trefoil and scabious. Sheep have traditionally grazed the downlands, but in recent years the grasslands have been under threat from intensive farming. The well-watered pastures on the Ferruginous Sands and Gault Clay support dairying, while livestock and arable farming can be seen on the Palaeogene sands and clays in the northern part of the island.

However, traditional agricultural products are difficult to market on the mainland due to high transport costs, and so horticulture involving the cultivation of higher-priced specialist crops that can better bear the costs of distribution has increased considerably. The growing of salad crops such as tomatoes, lettuce and cucumbers often takes place under glass or polythene tunnels. There is a concentration of glasshouses for market gardening on Hale Common [SZ 550840] in the Upper Eastern Yar valley, where fertile brickearth overlies the terrace gravels. Garlic is another crop that is cultivated, particularly

View from St Catherine's Down across productive agricultural land on the Ferruginous Sands around Chale.

around Newchurch [SZ 551859], and even exported to France! Lavender is grown on the gently undulating sandy clays of the Hamstead Beds at Staplehurst Grange near Newport. The long growing season on the island encourages horticultural enterprises (market gardening, nurseries and garden centres) in sheltered locations.

Viticulture has developed in recent decades and there are vineyards at Wellow near Yarmouth, Smallbrook near Ryde and Adgestone near Sandown. The latter is ideally sited on well-drained calcareous soils washed down from the slopes of the nearby Brading Down, and it also benefits from an almost frost-free microclimate in the sheltered scarp foot vale. Vines were, of course, grown here by the Romans and so this is really a case of reintroducing an ancient form of cultivation.

Despite these specialized horticultural innovations, the present gross annual value of all agricultural output represents only around 2.5 per cent of the total value of the Isle of Wight output, compared with 19.5 per cent for industrial activity and 78 per cent for the service sector.

Marshes and Wetlands

There are several nature reserves that have been developed in recent years. Alverstone Mead [SZ 580855] is an area of water meadows adjacent to the Eastern Yar, its ancient ditch drainage system attracting a range of wildfowl and wetland flora. Willow and alder carr is found along the southern boundary of the wetlands, which give way to coppiced woodlands on the higher ground.

Brading Marshes, behind Bembridge Harbour, is a nature reserve managed by the RSPB. It is drained by the Eastern Yar and consists of reed beds, lagoons, drainage ditches and water meadows bordered by ancient woodland. The reserve attracts winter migrating wading birds, and in spring and summer there are marsh harriers and peregrines, heron and shelduck.

Newtown Harbour is a designated National Nature Reserve administered by the National Trust. There are a range of different habitats here, including salt marsh, tidal creeks, wildflower meadows and secluded woodlands. Walter's Copse is home to red

Saltmarsh on the tidal flats on the western River Yar. Note the sluggish meandering river utilised by sailboats.

squirrels, and a variety of birds can be seen on the reserve, including lapwing, brent geese, curlew, wigeon and dunlin. Additionally, the wide, sheltered Newtown estuary offers a safe anchorage for sailing boats. The former gravel pits on the Yar river terraces between Horringford Bridge and Godshill provide small reservoirs, lakes and ponds for water fowl and freshwater fish.

Forestry

Woodlands cover the second largest acreage after agricultural land, and the larger ones are managed by the Forestry Commission. Parkhurst Forest, to the northwest of Newport, is a SSSI consisting of ancient deciduous woodland, relict heathland and plantation. The large forest is a haven for wildlife, particularly the red squirrel, which thrives on the island due to being isolated from the mainland and competition from its ubiquitous grey cousin. There are numerous woodland walks, bridle paths, cycle tracks and picnic areas.

The Forestry Commission was set up after World War I to develop plantations to provide a supply of homegrown timber. Coniferous plantations composed of serried ranks of pine trees were commonplace, but in more recent years there has been an increase in mixed woodland, which is more favourable to wildlife. Public access has also been increased as the recreational potential of forests has been realized. Brighstone Forest, in the southwest, is situated on rolling chalk country across Newbarn Down, Brighstone Down and Westover Down. It consists predominately of broadleaved deciduous trees, with beech forming a large proportion of the woodland. There are waymarked walks for visitors and access is from the National Trust Jubilee car park at SZ 419845 on the minor road from Brighstone to Calbourne.

The gardens of Osborne House.

Recreation

Tourism has made a big impact on the rural economy of the Isle of Wight in recent decades. The National Trust is a major landowner and manages some of the most scenic areas of river estuaries, chalk downland, and coastal cliffs, including the Newtown estuary, Bembridge Harbour, St Boniface Down, Tennyson Down and the Needles headland. It also owns several properties that are tourist attractions, such as Mottistone Manor gardens, Bembridge windmill, Newtown Old Town Hall and the Needles Batteries.

However, English Heritage manages the largest estate on the island – Osborne House, near East Cowes. This was built as a summer residence and retreat for Queen Victoria and Prince Albert. It was designed in the Italian Renaissance style and completed in 1851.The house stands in attractive landscaped grounds and is one of the island's main tourist draws. English Heritage also administers Carisbrooke Castle, which is another popular attraction on the outskirts of Newport.

Several country parks provide areas of outdoor recreation that are ideal for day visits to enjoy woodland walks, picnics and children's adventure activities. Fort Victoria Country

Park near Yarmouth contains the remains of a Victorian fort with fine views across the Solent, coastal walks and beautiful woodlands with nature trails. An exhibition on local shipwrecks and archaeology and a planetarium are also on site. Several theme parks attract thousands of visitors, especially during the summer months. The Needles Landmark Attraction at Alum Bay, the Blackgang Chine Amusement Park and Robin Hill Adventure Park near Downend are the principal enterprises, established in rural locations where sufficient land is available for large-scale tourist development.

There are numerous footpaths and cycle routes for both hikers and cyclists. The long-distance coastal path is 113km long and, apart from detours around the MOD range near Newtown and the Crown property at the Osborne estate, the route stays close to the coast. There is also the Vectis Trail, which crosses the island from Yarmouth to Brading via Carisbrooke and forms part of the E9 long-distance European pathway along the south coast of England.

For cyclists, there is an established round-the-island route about 92km in length, and two National Cycle Network routes: Number 23 from Cowes to Newport and Sandown along the former railway track, and Number 22 from Ryde to Wroxall and Newport. In addition, there are cycleways linking Merstone, Wroxall and Sandown, known as the Red Squirrel Trail, and between Freshwater and Yarmouth.

Camping and caravan sites are important for tourism and these range from small family-run farm sites with basic facilities to commercial holiday parks with swimming pools, shops and modern facilities. The Orchards Holiday Park near Newbridge is an example of a large commercial camping site, while the Whitecliff Bay Holiday Park offers top-of- the-range chalets and static caravans. Some small farms operate seasonal camping sites or bed-and-breakfast facilities as a way of diversifying their economy and bringing in additional income.

Recreational fishing is a popular holiday activity on the island. There are numerous ponds, lakes and streams for freshwater angling stocked with trout, carp, perch and other species. Saltwater angling is also popular, with inshore catches of bass, ray and bream taken by boats sailing out of Bembridge and Yarmouth.

The sandy beach at Compton Bay provides an ideal destination for holidaymakers.

CHAPTER 12

Impact of Tourism on the Countryside

The Isle of Wight is to many people a 'jewel in the crown' of the landscapes of southern Britain. It is scenically beautiful with undulating topography underlain by varied geological strata. From rolling chalk downs to woodland glades, landslipped cliffs, secluded estuaries, and sandy beaches, the island is a honeypot for tourists. It is not surprising that the local economy depends more on tourism than agriculture and industry combined. Much of the island is geared to holidaymakers, including ferry ports, hotels and guest houses, camping and caravanning sites, historic venues and sailing marinas, plus numerous walking and cycling routes.

Tourism has a positive impact in that it provides jobs for local people and encourages investment in local businesses. It also helps to preserve rural services like village shops, post offices and bus services. There is increased demand for local food production and the development of rural crafts is stimulated. Seaside towns that stagnated when package foreign holidays became fashionable in the late twentieth century have since been revitalized by more recent trends in 'staycation' holidays.

There are, however, also many negative aspects to tourism, particularly when the number of visitors exceeds the capacity of the land to sustain them – in other words, when a tourist attraction becomes so popular that it is damaged by the impact of too many visitors. This may, for example, lead to traffic congestion on narrow country roads or it cause erosion of footpaths by excessive numbers of hikers. Another negative factor is litter, some of which is flotsam brought in on the tide, which can be a serious problem on holiday beaches. In rural areas, careless visitors with dogs off the lead can disturb livestock or vandalize gates and fences. Everyday goods can be more expensive

The cliffs of Alum Bay are popular with visitors who can see the vertical beds of sands that are coloured to reds, yellows and browns by their various iron oxides. Sand collected from fallen blocks is used to make colourful souvenir ornaments.

A rainbow sky: looking across the 'Wealden coast' towards the chalk cliffs of Freshwater.

Chalk downland in its typical form with a rich grass cover on gently rounded terrain, and with a valley that is dry because most of the drainage is underground in the permeable chalk.

in the tourist season and the demand for holiday homes can make housing too expensive for local people. Although employment is created by tourism, most jobs are seasonal and low paid.

A Protective Framework

One of the most important initiatives designed to protect the rural environment was the creation of Areas of Outstanding Natural Beauty (AONB) in 1963, and this designation covered almost half of the Isle of Wight. The area is not continuous but is made up of five discrete parcels of land across the island. The Countryside Agency (now Natural England) stated in 2001: 'The primary purpose of designation is the conservation and enhancement of natural beauty, which includes wildlife and cultural heritage as well as scenery'. The AONB plan sets out several distinctive landscape character types, which include:

Chalk downlands, an open landscape of grassland or large fields, few mature trees and sparse hedgerows

Traditional enclosed pasture, found on the heavier clay soils of the Hamstead Beds in the north of the island. This is a landscape of lush

continued on p.106

Itinerary 20: A Heritage Coast – Newtown Bay to Gurnard

Walking distance: Porchfield to Gurnard 6km

The coast between Newtown Bay and Gurnard Bay in the northwest of the Isle of Wight provides a good example of an unspoilt and sparsely populated area of conservation. The Newtown estuary is a branching drowned river system protected by paired spits at its mouth. Newtown Harbour National Nature Reserve is managed by the National Trust so that its natural environment of tidal mudflats, coastal marshes, water meadows and ancient woodland are protected. The whole area is open to the public via a network of pathways that extend out across the mudflats to Clamerkin Lake, the eastern arm of the Newtown River. Other tracks run through Walter's Copse, now a haven for wildlife, which is recorded as being open meadow during medieval times. There are car parks at Shalfleet and Newtown, providing access for visitors wishing to explore this quiet backwater on foot.

The area immediately to the east of the nature reserve is inaccessible due to MOD restrictions; the boundary of the danger area is marked on the Ordnance Survey map. The coast path circumvents the military area and so it is best to park in Porchfield, from where you can walk to the shoreline. The most direct way is to take the path through Elmsworth Farm and alongside Burnt Wood, and thence along the slumping cliffs of Bembridge Marls towards Thorness Bay. The shore platform is formed of Bembridge Limestone but is difficult to walk on due to slumped mud and seaweed. However, below Thorness Wood [SZ 449933], there is a good section of the upper part of the Bembridge Marls with bivalves concentrated in shell bands and muds containing freshwater gastropods.

An alternative route from Porchfield is to join the coast path at SZ451917, then walk through Thorness Bay Holiday Park, reaching the coast just west of Little

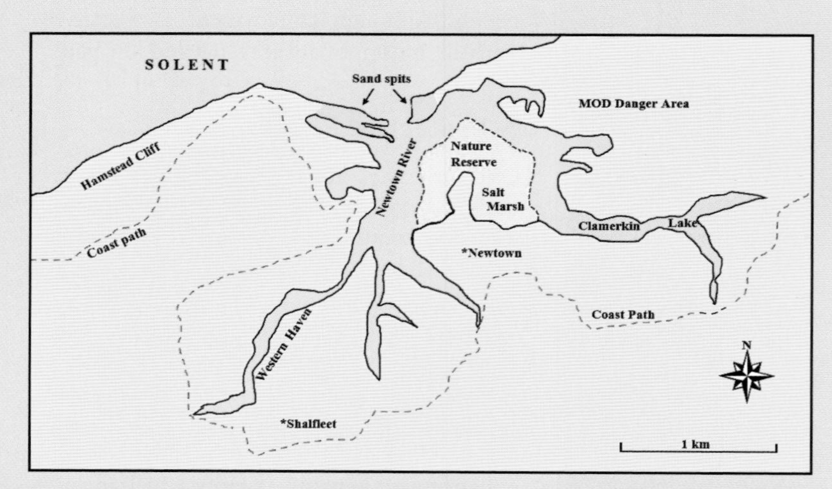

Map of the Newtown estuary and the surrounding salt marsh, which form a nature reserve. The estuary is a drowned river mouth, or ria.

Newtown National Nature Reserve. Tidal creeks and coastal salt marshes provide a haven for a variety of wildlife.

Thorness Marsh. A footbridge crosses the Thorness stream, after which the path follows the edge of the low, slumping terraced cliffs that rise towards Gurnard Ledge on the north side of Thorness Bay.

Bembridge Insect Bed

Here the Bembridge Limestone dips gently southwest and extends seawards as a prominent reef. Just south of Gurnard Ledge, at SZ 464945, the base of the Bembridge Marls can be seen resting on the Bembridge Limestone, and about a metre above this junction is the famous Bembridge Insect Bed, containing thin bands of limestone. This site is considered to be a lagerstatte, that is a sedimentary deposit that exhibits extraordinary fossils with exceptional preservation. The bed has yielded over 200 species of insect including beetles, spiders and dragonflies and the leaves of palm, fig and cinnamon, showing that the climate during the early Oligocene was subtropical.

In the Bembridge Marls above the insect bed there are numerous shelly bands where bivalves are concentrated, and near the top of the succession are assemblages of freshwater gastropods. There is a capping of Hamstead Beds at the top of the cliff. North of Gurnard Ledge, the slumping cliffs decline but the Bembridge Limestone shore platform can be traced as far as the mouth of the Gurnard Luck Valley. Here the road reaches shoreline and there is parking space at the end of Marsh Road, where you can arrange for return transport to Porchfield if required.

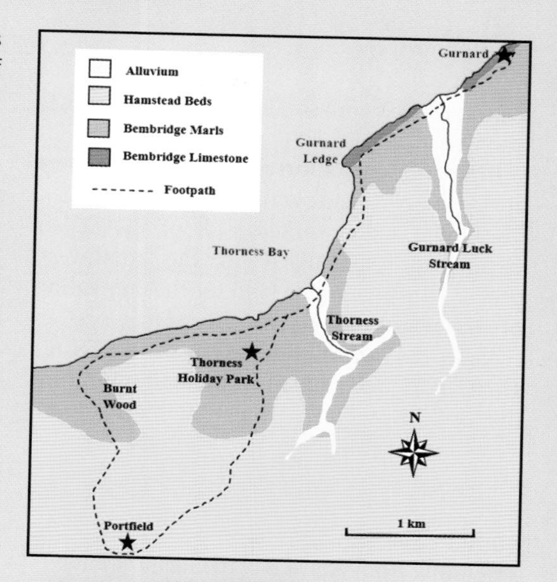

TOP: **Map of Thorness Bay and Gurnard Bay, with the coastline formed of slumping Bembridge Marls.**

RIGHT: **A minor anticline in the Bembridge Marls in Thorness Bay. The hammer is beside a normal fault.**

BELOW: **Managed by the National Trust, the Newtown estuary offers sheltered anchorage for sailing boats.**

continued from p.103

green pastures with small fields, hedges, copses and woodlands.

Intensive agricultural land, where there are large open fields with few trees or hedges. The light, easily worked soils of the Lower Greensand, particularly in the southwest, provide the best arable land for the cultivation of cereals and root crops.

Northern woodlands, areas of coniferous plantations and mixed woodland found in the northwest and in the hinterland of the Wootton estuary

Other distinctive areas include the harbours and creeks of the north coast, the Undercliff on the south coast and the slumped cliffs extending from Alum Bay to Gurnard Bay.

Heritage Coast

The Hamstead Heritage Coast lies within this last area and has been designated to reflect its unspoilt character and its importance for nature conservation. The Tennyson Heritage Coast extends from Totland Bay past the Needles, down the southwest coast to Steephill Cove in Ventnor. Although sharing the aims of the AONBs, the Heritage Coast status is designed to provide both protection and sensitive management to areas of natural beauty that are enjoyed by the public. This involves improving and extending appropriate recreational, educational, tourist and sporting opportunities where they do not conflict with the conservation of the resource.

The Heritage Coast management extends 2km out to sea, but in recent years three Marine Conservation Zones (MCZ) have been established around the Isle of Wight in order to protect marine habitats and species. The inshore waters between the Needles and Fort Albert form a conservation area that supports significant colonies of seagrass and rare seaweeds. Between Yarmouth and Cowes is an MCZ that hosts diverse submarine communities, including sea squirts, sponges and porcelain crabs. Finally, there is the Bembridge MCZ, where the shallow waters contain large seagrass beds that provide a habitat for species such as cuttlefish, stalked jellyfish and seahorses.

Map of the Areas of Outstanding Natural Beauty (AONB), Heritage Coasts and Marine Conservation Zones (MCZ) on and around the Isle of Wight.

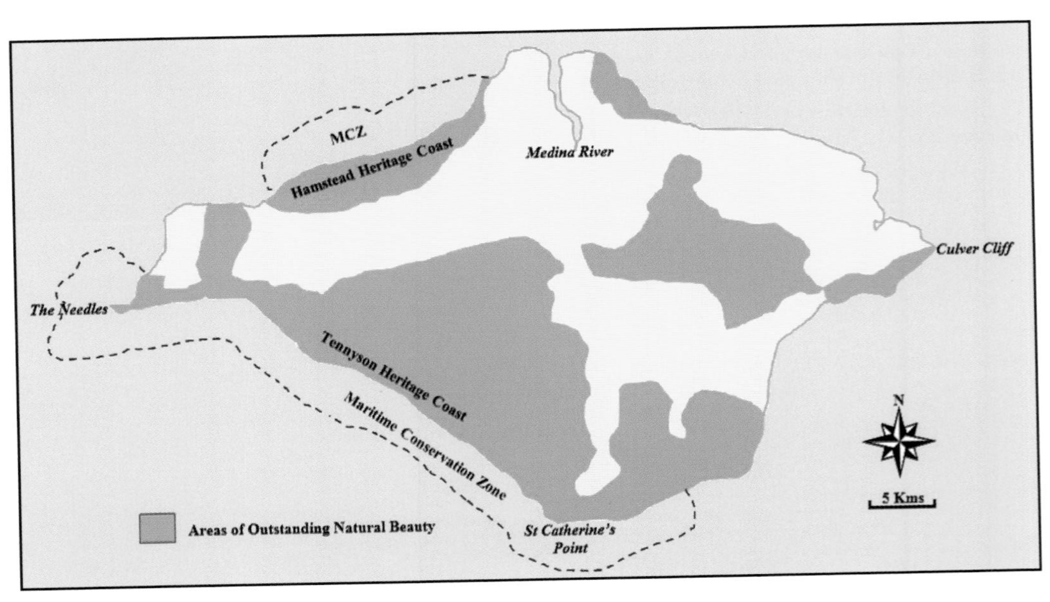

ABOVE: **The view across Freshwater Bay to the delightful holiday resort with the same name, looking across the long shingle beach towards the western headland with its chalk cliffs.**

BELOW: **A party of field geologists from The Open University examine dinosaur footprints in the Wessex Formation on the foreshore at Hanover Point.**

Further Reading

Bird, E., *The Shaping of the Isle of Wight with an Excursion Guide* (Ex Libris Press, 1997)

Gale, A., *The Isle of Wight*, Geologists' Association Guide No. 60 (6th edition, 2019)

Hobson, P., 'The geological history of the Isle of Wight: an overview of the 'diamond in Britain's geological crown' (*Proceedings of the Geologists' Association*, 122, 2011, p. 745–63)

Hughes, J. Cecil, *The Geological Story of the Isle of Wight* (Edward Stamford, 1922)

Isle of Wight AONB Partnership, *Isle of Wight, Area of Outstanding Natural Beauty: Management Plan 2019–2024* (wightaonb.org.uk)

Lott, G.K., 'The use of local stone in the buildings of the Isle of Wight' (*Proceedings of the Geologists' Association*, 122, 2011, p. 923–32)

Melville, R.V. and Freshney, E.C., *British Regional Geology: The Hampshire Basin and Adjoining Areas* (HMSO, 1982)

Natural History Museum, *British Fossils: Cenozoic* (Natural History Museum, 2017)

Natural History Museum, *British Fossils: Mesozoic* (Natural History Museum, 2013)

Osborne White, H.J., *A Short Account of the Geology of the Isle of Wight*, Memoirs of the Geological Survey of Great Britain, 1921 (HMSO, reprinted 1990)

Toghill, P., *The Geology of Britain* (Swan Hill Press, 2000)

Trueman, A.E., *Geology and Scenery in England and Wales* (Penguin Books, 1971)

Maps

British Geological Survey, 1:50,000 Special Sheet, Isle of Wight (2013)

Ordnance Survey Explorer Series 1: 25,000, Sheet OL29, Isle of Wight (2015)

Ordnance Survey Landranger Series 1:50,000, Sheet 196, Solent and the Isle of Wight (2010)

Glossary

Alluvial fan A fan-shaped accumulation of sediment deposited where a river gradient decreases rapidly at the foot of a mountain range

Anticline Arch-shaped upfold in strata, with the oldest rocks at the core

AONB Area of Outstanding Natural Beauty

Aquifer Sedimentary bed of porous and permeable rock that contains significant amounts of water

Ashlar Block of finely dressed building stone with rectilinear edges

BGS British Geological Survey

Bioclastic limestone Grains in limestone that are formed of organic remains such as shell fragments

Bioturbation Churning of sediment by burrowing organisms such as arthropods and bivalves

Boulder clay Glacial deposit of clay containing ill-sorted sub-angular fragments laid down beneath a glacier or ice sheet. The term glacial till is now more commonly used.

BP Abbreviation meaning 'before present'; by convention, the 'present' is 1950

Brachiopod Sessile marine invertebrate with bivalve shell attached to rocks by a stalk

Braided river River that splits into many channels separated by sand and gravel bars

Breccia Coarse-grained fragmental rock containing angular clasts set in a finer-grained matrix

Calcrete Carbonate horizon formed by evaporation of calcium minerals in a semi-arid environment.

Carr Waterlogged swamp woodland; wet peat soils colonized by willow, birch and alder with sedge and reed undergrowth

Chert Microcrystalline form of silica that forms nodules and bands. Usually of biological origin, derived from siliceous ooze on the seafloor; also formed as a chemical precipitate

Clast Fragment or particle of rock varying in size from silt-sized grains to boulders

Clastic rocks Composed of fragments, or clasts, of pre-existing minerals and rock, as for example, in mudstones and sandstones

Concretion Rounded rock body that has been cemented within a sediment at an early stage during diagenesis

Conglomerate Coarse-grained fragmental rock containing rounded clasts set in a finer-grained matrix

Cross-stratification Layers of sediment that are inclined at an angle to the surface on which the sediment was laid down. Usually seen in shallow marine, deltaic or wind-blown deposits. Also known as cross-bedding

Curvette Large basin of deposition produced by subsidence

Diagenesis The processes of physical and chemical changes that take place during the conversion of sediment into sedimentary rock

Dip The inclination of a planar surface measured in the vertical plane at right angles to the strike

Distributary channels Branching channels into which the main river splits as it crosses a delta plain

Erratic Rock boulder that has been transported by ice and deposited far from its original location

Eustatic Relating to global changes in sea level caused either by the expansion or melting of ice caps or by tectonic movements

Ferruginous Having an abundance of iron minerals, such as haematite or limonite

Fluvial Relating to the processes and actions of rivers

Glauconite A complex iron potassium silicate with a characteristic greenish colour formed under marine conditions. It is one of the main components of greensand

Graded bedding Sedimentary layers that show a vertical gradation in grain size

Haematite Iron oxide, Fe_2O_3, commonly forms a cement in red sandstones. A major ore of iron. Varieties include kidney ore and specularite.

Head Poorly sorted angular rock debris moved downslope by solifluction under periglacial conditions

Heterolithic beds Fine interbedding of sand and mud produced by variable current flow

Interfluve A ridge of high land separating adjacent valleys containing streams flowing in the same direction

Isostasy State of balance where the Earth's crust rests in equilibrium on the underlying mantle

Lag gravel A residual accumulation of gravels at the base of a river channel

Law of Superposition When strata are deposited sequentially so that each layer is younger than the layer beneath it

Lithification Process of converting loose grains into rock through compaction and cementation

Loess Silt-sized sediment that is formed by the accumulation of wind-blown dust

Longshore drift The transport of sediment along the shore in one direction as the result of oblique wave action

Mass movement Transfer of surface material downslope; includes rock fall, rock slide, mudflow and soil creep

Matrix Fine-grained material separating the clasts in a sedimentary rock

MCZ Marine Conservation Zone

Monocline Fold with one steep limb and one almost horizontal limb.

Normal fault A fault in which the displacement of the hanging wall is downwards relative to the footwall. This type of fault is under tension, so the rocks are pulled apart.

OD Abbreviation for Ordnance Datum; usually mean sea level, used as a base for deriving altitudes on OS maps

Orogenesis Mountain-building process in which sediments are compressed, uplifted and folded as plates converge

Palaeocurrent A fluvial or marine current from which sediment was deposited in the past

Periglacial Area on the margins of an ice sheet where permafrost exists

Permeability Ability of a rock to allow a fluid to pass through either the pore spaces between the grains or through fissures in the rock

Point bar A depositional feature formed of sediment that accumulates on the inside bend of a meandering river

Porosity Ability of a rock to contain fluids within the pore spaces between the grains

Raised beach Former beach that is now several metres above present sea level as a result of either a fall in sea level or uplift of the land.

Rendzina Calcareous soil with thin humus layer developed on chalk downland.

Reverse fault A fault in which the displacement of the hanging wall is upwards relative to the footwall. This type of fault is under compression, so the rocks are squeezed together.

Septarian nodule Large concretion of clay ironstone characterized by radiating internal mineral-filled cracks.

Solifluction Slow downslope movement of water-saturated surface material that often takes place when permafrost melts in summer

SSSI Site of Special Scientific Interest

Strike The direction of a horizontal straight line constructed on an inclined planar surface at right angles to the direction of true dip

Suspension deposit Fine particles that stay in suspension in water until the current velocity slows sufficiently for the material to be deposited

Syncline U-shaped downfold in strata, with the youngest rocks at the core

Trace fossil Structure in sediment produced by an ancient organism, for example a burrow, trail, footprint or faecal material

Unconformity Break in the stratigraphical record, usually marked by an erosion surface

Wave-cut platform Rocky foreshore planed by marine erosion at the base of receding cliffs

Zone fossil Distinctive and abundant fossil chosen to represent a particular biostratigraphical zone.

Index